Effective Methods for Software Engineering

Effective Methods for Software Engineering

Boyd L. Summers

CRC Press is an imprint of the
Taylor & Francis Group, an **informa** business
AN AUERBACH BOOK

First edition published 2020
by CRC Press
Taylor & Francis Group
6000 Broken Sound Parkway NW, Suite 300
Boca Raton, FL 33487-2742

First issued in paperback 2022

ISBN 13: 978-1-03-247435-9 (pbk)
ISBN 13: 978-0-367-45838-6 (hbk)
ISBN 13: 978-1-003-02566-5 (ebk)

DOI: 10.1201/9781003025665

Publisher's Note
The publisher has gone to great lengths to ensure the quality of this reprint but points out that some imperfections in the original copies may be apparent.

Library of Congress Cataloging-in-Publication Data

Names: Summers, Boyd L., author.
Title: Effective methods for software engineering / by Boyd Summers.
Description: Boca Raton : CRC Press, 2020. | Includes bibliographical references and index. | Summary: "This book presents methods for analyzing software engineering problems, designing solutions, coding effectively, and testing software"— Provided by publisher.
Identifiers: LCCN 2019058121 (print) | LCCN 2019058122 (ebook) | ISBN 9780367458386 (hardback) | ISBN 9781003025665 (ebook)
Subjects: LCSH: Software engineering.
Classification: LCC QA76.758 .S857 2020 (print) | LCC QA76.758 (ebook) | DDC 005.1—dc23
LC record available at https://lccn.loc.gov/2019058121
LC ebook record available at https://lccn.loc.gov/2019058122

Visit the Taylor & Francis Web site at
http://www.taylorandfrancis.com

and the CRC Press Web site at
http://www.crcpress.com

Contents

List of Figures

List of Tables

Preface

For many years, I always wanted to write a book related to software engineering. My new book is titled *Effective Methods for Software Engineering.* Software engineering activities ensure effective methods to be performed for accomplishment of projects in companies, institutions, military programs, and successful businesses. I will also provide the importance of software programming, software configuration management (SCM), Rational ClearCase and ClearQuest tools, software quality engineering (SQE), systems engineering (SE), software applications, verification and validation, and describe how software engineering is involved and followed in software programming development framework and release to customers. My previous books *Software Engineering Reviews and Audits*, *Effective Methods for Software and Systems Integration*, and *Effective Processes for Quality Assurance* provided the framework and detailed requirements for all software engineering activities.

Software Engineering Reviews and Audits

Software Engineering Reviews and Audits will improve individual and company efforts in maintaining a professional setting where quality is improved for increased profit, cost reduction, control, and service improvement. Performing reviews and audits for quality assurance in companies and institutions is successful in ensuring compliance with specified contracts and customer satisfaction.

Effective Methods for Software and Systems Integration

In order to develop, operate, and maintain software and systems integration capabilities inside work product facilities and companies, there must be a major discipline that supports to completely understand the entire software life cycle (i.e., planning, systems, requirements, software design, builds, installations, integration, subcontractors, quality, and delivery to customers).

Effective Processes for Quality Assurance

The book provides an understanding of the importance of day-to-day activities of quality assurance to advocate a culture that supports commitment to customer integrity.

The primary purpose is to increase the implementation of *Effective Processes for Quality Assurance* in companies and institutions to increase communication, knowledge, and visibility in their operations. This book provides information to convey the methods for quality assurance to be more effective in organizations, which can benefit from adopting these processes.

Summary

It is critical to understand and implement the disciplines of *Effective Methods for Software Engineering* prior to deliveries of software product development to foster successful business environments. Chapters in this book can help ensure software engineering employees are trained to perform reviews, audits, and evaluations that can provide effective oversight of software engineering activities.

Acknowledgments

The opportunity to be an Hercules Missile Defense and Boeing Defense and Space employee, and consultant for companies, institutions, military programs, and successful businesses made me a team player and to provide software engineering expertise. Currently, I am a Software and Quality Assurance Technology Consultant supporting all software engineering and quality assurance activities and providing expertise and solutions for software engineering capabilities for businesses and institutions. To ask questions for current and future software engineering solutions, contact me through the email bl.summers.consulting.llc@gmail.com; all important information related to this topic will be provided, discussed, and fixed.

Author

Boyd L. Summers is currently a retired software engineer for The Boeing Company and is a Software Technology and quality consultant for BL Summers Consulting LLC living in Florence, Arizona. He is the author of three software technology books titled *Software Engineering Reviews and Audits, Effective Methods for Software and Systems Integration*, and *Effective Processes for Quality Assurance.* He has provided written software articles to many software engineering journals and magazines. His topics of interest include *Software Engineering, Software Programming, Software Requirements, Software Test and Evaluation, Software Configuration Management, Software Quality Assurance, Quality Assurance, Audits, Reviews, and Software Engineering Evaluations.* He has applied processes in Agile, Lean, and Six-Sigma. He is a speaker and a board member of numerous software and quality control conferences in the United States and countries around the world.

BL Summers Consulting LLC
Software and Quality Assurance Solutions
7291 W. Sonoma Way
Florence, Arizona 85132
Phone: (206) 485-8618
E-mail: bl.summers.consulting.llc@gmail.com

Chapter 1

Introduction

1.1 The Role for Software Engineering

The role of software engineering is to introduce the notion of software as a product designed and built by software engineers using effective methods. Software is important because it is used by a great many people in companies and institutions. Software engineers as programmers should have a moral and ethical responsibility to ensure that the software and design they provide do not cause any serious problems. Software engineers tend to be concerned with the technical elegance of their software products and tools, whereas customers tend to be concerned only with whether a software product meets their needs and is easy and ready to use.

For two years, I was a software programmer for Hercules Missile Defense, where my responsibility was to make sure that the software that is available for implementation and use was programmed and performed according to requirements.

In order to develop, operate, and maintain software engineering capabilities, a major discipline in supporting software programming, design, builds, and delivery to customers should be completely understood. A critical understanding of and a right approach for the methods used for software engineering is necessary to enhance the process and achieve effective programming results.

The right disciplines are identified in Figure 1.1.

It is more than a discussion and a way of life for software engineering to be important related to the standards and requirements for programming and design. Software engineering must comply with these standards and requirements during several stages of the development process.

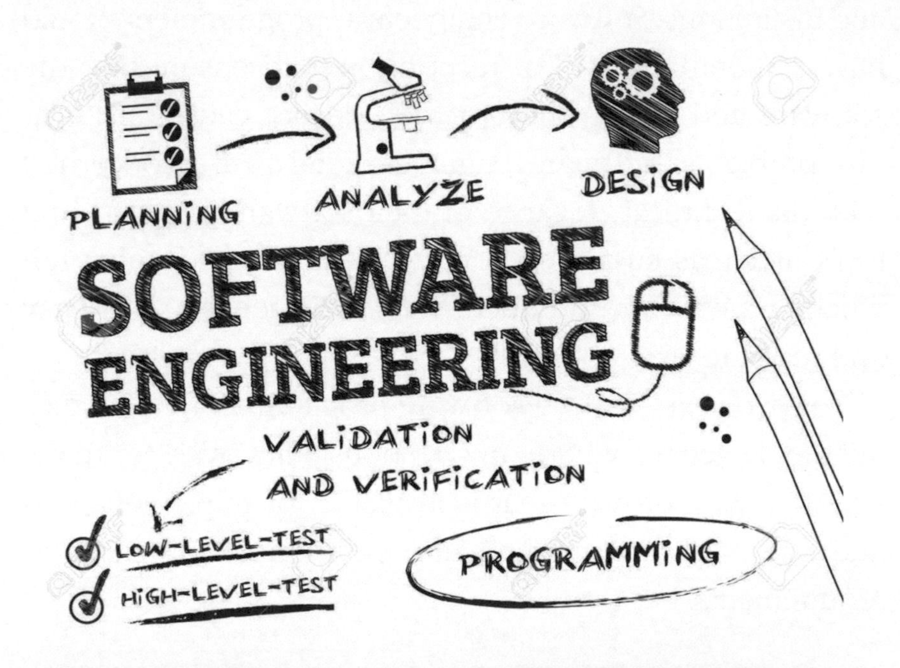

Figure 1.1 Right approach for software engineering.

The standards and requirements adopt and deploy the following:

- Understand the basic concepts, standards, and requirements of software engineering.
- Select appropriate techniques for programming and designing.
- Effectively use software engineering tools and applications.
- Create specifications to comply with the software standards and requirements.
- Utilize various standards and requirement techniques to identify defects.
- Manage changes to standards and requirements.

1.2 Software Engineering Process

This section discusses the concept of a software process framework and provides a brief overview of the software engineering Capability Maturity Model Integration (CMMI). It is important to emphasize that CMMI does specify which process model needs to be used for a given project or organization. Rather, it provides a means of assessing how well company and institutional processes allow them to complete and manage and program new software projects.

Software engineering divides the process of software development into distinct phases to improve programming and designing, which is known as "software development life cycle".

Most modern software engineering processes can be described as Agile. Other methodologies include waterfall, prototyping, iterative and incremental development, spiral development, rapid application development, and extreme programming.

Software engineering considers "life cycle model" as a more generic term for a category of methodologies and "software engineering development process" as a more specific term to refer to a process chosen by a company or an institution. There are many specific software engineering processes that fit the life cycle model.

Software engineering is the process of envisioning and defining software solutions to one or more sets of problems. One of the main components of software engineering requirement (SER) is part of the software development process that lists specifications used in software engineering. If the software is semi-automated for software engineering, it can help determine those specifications. If the software is completely automated, software engineering may be as simple as describing a planned sequence of events. In either case, a software plan is usually the product of the software engineering and design. Furthermore, software engineering may either be platform-independent or be platform-specific depending on the availability of the technology used for programming processes.

1.3 Software Engineering Planning

It is very important for software engineering programmers to have some basic knowledge of planning and performing software projects. Software engineering planning involves estimating how much time, effort, and resources are required to build a software product. There is enough information provided in the text that allows programmers to estimate their own projects and write their own planning documents. Software engineering programmers should be assigned the task of writing planning documents using appropriate templates as part of their coursework roles.

First, management and software engineering programmers must prepare for planning by identifying contracts and requirements evaluated during a specific period for companies, institutions, military programs, or successful businesses. Then, the identified contracts and requirements require the right criteria derived from associated software plans, documents, procedures, and work instructions.

1.4 Software Applications

Software applications are different from the artifacts produced in most other engineering disciplines. Opportunities for replication without customization are not very common. Software may deteriorate, but it does not wear out. The chief reason for software engineering is that many changes are made to software products over their lifetime. When changes are made, defects may be inadvertently introduced to portions of the software that interact with the portion that was changed. The most popular tools used by the military and defense programs are Rational ClearCase and ClearQuest. I implemented these tools to many Boeing Defense and Space programs to be used by software engineers, software configuration managers, and testers, and I will explain the use of these two tools in Chapter 6.

1.5 Software Engineering Programming

Software engineering programming encompasses preprogramming compilers that effect the translation of languages from a variety of editors to software codes to be used by software engineers in their tools. Software engineers that do programming can be important for software design and delivery. Programming implements design characteristics for the programming language used. There are two design topics used:

- Data design.
- Procedural design.

The strategy for software engineering programming is illustrated in Figure 1.2.

Economies of the developed software are dependent on software engineering. In recent days, more and more

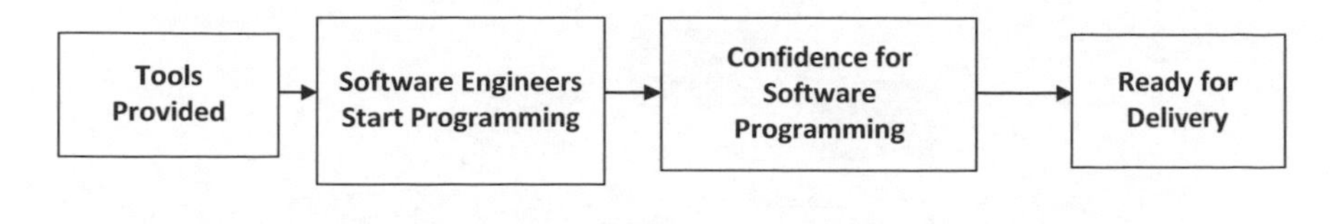

Figure 1.2 Strategy for software engineering programming.

systems are software-controlled and concerned with the methods and tools used for professional software engineering programming and development.

The elements of software engineering are as follows:

- Software requirements and software design.
- Software testing and maintenance.
- Software configuration management (SCM) builds, plans, and documents.
- Software engineering management.
- Software engineering plans, processes, and procedures.
- Software engineering tools and methods.
- Software quality engineering reviews and buy-offs.

> When poor Software Engineering methods and designs are not effective; that can lead to major problems with management, team members, employees and customers.
>
> *Boyd L. Summers*

Summary

Software engineering is a detailed study of designing, programming, developing, and maintaining software products. It is introduced to address the issues that arise due to low

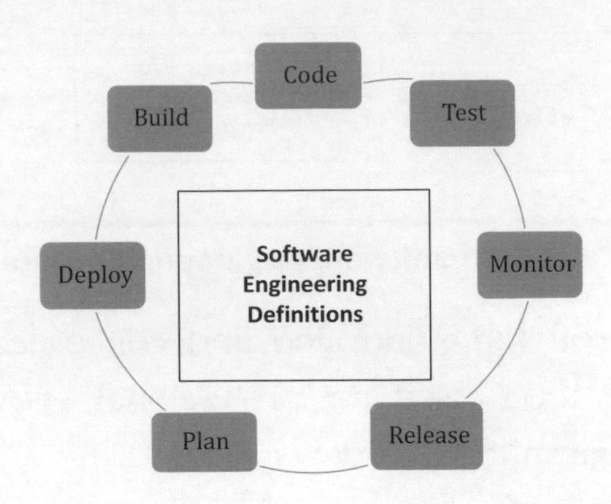

Figure 1.3 Software engineering definitions.

quality of software products. Problems arise when development of a software product exceeds its timeline and/or it has reduced levels of quality. Software engineering ensures that the application is built consistently, correctly, on time, and according to requirements. The demand of software engineering also emerges to cater to the immense rate of change in user requirements and environment on which applications are to be working (Figure 1.3).

Further Reading

Long, L. N., lnl@psu.edu 2005. www.personal.psu.edu/lnl.
Pressmen, R. S., *Software Engineering Book: A Beginner Guide.* 1988. ISBN 0-07-050790-1. www.ebay.com/itm/143051317925.

Chapter 2

Software Engineering Peer Reviews

Software engineering peer reviews provide disciplined practices for detecting and correcting defects in software programming, thus preventing any issues and problems in their operations. Peer reviews identify and fix problems while performing software engineering activities in early stages of software development. Thus, for example, any requirements-related problems are identified and fixed during the analysis of requirements, rather than doing it during the stages of programming and testing.

2.1 Software Testing

Software testing involves execution of a software programming to evaluate the interest of a software product. It indicates to which the software and systems are tested:

- Accomplish the software programming that guides for design and development.

- Always respond correctly to all kinds of software programming and systems under test.
- Perform all functions within an acceptable time for software engineering development.
- Achieve the results of all software and systems under test.

As the number of possible tests for even simple software engineering components can be infinite, a sound software testing strategy is used to select tests that are useable for available time and resources.

Software testing is typically used to execute software programming with the intent of finding software engineering errors and defects. It is an iterative process for fixing issues in an existing software, and it can even create new software with required capabilities.

Software testing can provide information about the quality of a software and the risk of its failure to software engineering programmers. It can be conducted as soon as an executable software exists and determines when and how testing is conducted in processes after system requirements have been defined and implemented in test programs. In contrast, under an Agile approach, requirements, software engineering programming, and testing are often done together.

2.2 Software Programming Support

Software programming support includes software engineers and software quality engineers to help prevent software programming bugs that are needed to be fixed before delivery to software configuration management (SCM) for builds. Although a code written by a software programmer is syntactically correct, it may fail in peer reviews and checks. Considering software engineering cost and performance, some software engineering languages deliberate and exclude bug problems at the expense of slower performance with a lower code.

I will give you an example using Java Programming that does not support implementations of some languages such as Pascal and scripting languages that have runtime arrays for debugging builds.

Tools used for the analysis of software engineering code help software engineering developers to spot potential problems by inspecting the program test and code beyond the software capabilities.

Although in general the problem of finding all software programming errors is a specification that is not solved, software programming tools exploit the fact that software programmers tend to make certain that there are fixes often when writing software coding.

Software programming tools monitor the performance of software development as it is occurring to detect problems and to to give assurance that code is working correctly. It is often to find where most of the time is taken by a piece of software code, and removal might cause the software code to be rewritten.

2.3 Peer Reviews in Software Engineering

Peer reviews in software engineering companies with experts are positioned at the top of the desirable development software engineering practices. Experience has shown that problem solving of defects is eliminated when processes incorporate peer reviews and that these peer reviews are effective for testing. It is therefore important for software engineers to perform peer reviews during software design projects.

It is important to incorporate peer reviews in software engineering to design projects for the purpose of presenting the experimental findings, lessons learned, possible challenges, and recommendations that may be used to promote learning and also the use of peer review activities for software engineering and computer engineering. The results show promising signs of using peer reviews in all software design projects.

2.4 Lessons Learned for Peer Reviews

Lessons learned for peer reviews in software engineering are related activities for software engineering programming and for teaching peer reviews.

I've had the opportunity to support software engineering peer reviews in software development organizational programs. A complete discussion about software engineering peer reviews needs to remember that a software engineering peer-reviewing process deals with three main objects:

- Software work product, software code, and design are needed to be reviewed.
- A software reviewer acts as observer and provides comments.
- A software engineer must get their software work revised by a peer review.

I have had many discussions supporting software engineering peer reviews as a quality engineer to ensure quality assurance processes are used. What really matters is that software engineering peer review is a useful process to improve the software product coding before an important release. I see the benefits of software engineering peer reviews as related to software quality assurance and control.

No matter how experienced software engineering developers are, a checklist should always be used. This checklist can be considered an organizational and program asset for projects where the checklist gets its initial definition. By definition, a checklist is a list of items to be checked and observed. I also recommend grouping the items to be checked by software engineering development. For example, groups of items documented for software peer reviews are stated as follows:

- Software programming items for use and implementation are reviewed.
- Items for software programming coding are performed consistently.

Software developers will always welcome a report template. Depending on the specific context of your organization, a report template can be viewed as either a project or an organizational asset. Your template should present to the reviewer a brief indication of concepts, scales, and possible values that will be used for evaluation purposes, irrespective of the case. For example, in your report template, you should be informing the reviewers what a "pass" or "fail" means.

If you are able to devote time for Lessons Learned for Peer Review Software Engineering activities, you can prepare for the peer review activity ahead of time by reviewing and having peer review feedback and ask clarification questions. Before the submission deadline for draft, inform software engineers that they will be performing peer review activities. Before this activity, remind software engineers to print off a copy of their reflections by adding the following announcements to learn peer review assignments as a reminder. You will share your draft reflection with your software engineers during the peer review activity.

To participate in the activity, software engineers must

- Submit the draft for deadlines.

- Have computers or other electronic devices to view software engineering programming.

Further Reading

O'Neil, D., 2001, *"Peer reviews" in Encyclopedia of Software Engineering*, New York: Wiley-Interscience.

Vahid, G., What we know about smells in software code test code. *IEEE Software*, 2019, Vol. 36, No 3, pp. 61–73. https://ieeexplore.ieee.org/xpl/RecentIssue.jsp?punumber=52.

Wiegers, K. E., 2002, *Peer Reviews in Software: A Practical Guide*, Reading, MA: Addison-Wesley.

Chapter 3

Systems Engineering Capabilities

System engineering ensures that all likely aspects of a project or a system are considered and integrated into programs. Software engineering is an interdisciplinary field of engineering that focuses on how to design and manage all systems of software life cycles. Systems engineering utilizes systems processes to organize important knowledge capabilities.

Systems engineering is an important field of software engineering and management that focuses on how to design and manage complex systems of software life cycles. The outcome of systems engineering activities can be used in software programming activities. The outcome of efforts is a combination of components that work in to collectively perform a useful software programming and design activities.

Systems engineering aspects such as requirements, reliability and coordination within software engineering teams, testing, evaluations, and other disciplines are important for successful system development and dealing with large and complex projects. Systems engineering deals with work processes, optimization methods, and risk management for all software programs and projects. It overlaps technical disciplines of software engineering, programs, and project management.

Systems engineering is a process of discovering effective methods to achieve high-quality outputs with minimum cost and time. This process begins by detecting the real problems that need to be resolved and finding solutions to these problems.

3.1 Scope of Systems Engineering

The scope of systems engineering covers all the activities performed by software engineers and management of an engineered system. These activities can be part of the systems engineering environment and functions. In the international activities, systems engineering top-level definitions approach and means to enable the realization of successful systems and software engineering development. The scope of systems engineering is shown in Figure 3.1, which shows the relationship between systems engineering and the software engineering life cycle.

Scope of systems engineering includes activities such as analyzing methods for production, testing, and operations for planning and analysis functions. These activities

Development Phasing
1. Baselines - Systems Engineering Management - Life Cycle Integration
2. Systems Engineering Process - Integrated Teaming - Life Cycle Planning

Figure 3.1 Scope of systems engineering.

involved in production and installation of software products that are used for programming and designing and for supporting the production environment.

3.2 Systems Engineering Concepts

The concepts of system engineering capabilities have been deemed relevant for many years. It is important to understand software engineering and the capabilities of making technical decisions. System engineering capabilities have some similarities with services and high activities.

The concept of service is to ensure that the capability and operations are directly related to the implementation of software engineering.

The purpose of this chapter is to summarize the major events of the systems engineering capability. The Systems Engineering Capability Maturity Model (SE-CMM) provides efficient methods for appraising the capability of programs and projects to reach required benchmark for achieving effective systems engineering functions. This chapter describes each step of an SE-CMM appraisal and provides guidance for the preparation and conduct of an appraisal.

Table 3.1 Systems Engineering Support Environment

1. Does the project or organization allocate adequate resources including team members for performing system engineering support process area on projects?
2. Does the projects have documented plans, standards, or procedures in place for performing system engineering?
3. Does your project or organization provide appropriate tools for system engineers supporting the performance of process areas on your project?
4. Does someone in the organization or project ensure that team members performing in process areas on projects have the appropriate skill or knowledge?

3.3 Systems Engineering Support

The process associated with systems engineering support environment process area is managed using the practices in Table 3.1.

3.4 Systems Engineers' Role

Systems engineers play an integral role in the success of programs, projects, and businesses in many industries. They lay the foundation for systems to begin conception, production, and realization of valuable operations. Their role includes identifying a problem based on the needs of software engineering and developing a solution which will constantly be evaluated throughout the process after its execution. Systems engineers monitor the performance of systems and continuously assess all stages of operations to ensure that there are no problems.

Systems engineers' role demands an ability to apply an interdisciplinary approach to ensure that technical procedures are translated into step-by-step activities for producing efficient and effective results. A systems engineer will often collaborate with a program and project manager along with software engineering teams to take the lead to facilitate the achievement of a successful system. They will also participate in each stage of the given system or process, from software design and development to validation and operation and risk assessment, often focusing on performance, testing, scheduling, and costs.

The type of work involved must pursue a career as a system engineer highly proficient in software-related topics. Many systems engineers should maintain constant communication with engineering teams, project managers, and other stakeholders in the programs and projects. Ultimately, aspiring systems engineers should have highly developed interpersonal skills.

Further Reading

Miller, T. R., Lt Col, USAF SEI Joint Program Office. This work is sponsored by the U.S. Department of Defense. Copyright © 6/3/96 by Carnegie Mellon University.

Chapter 4

Systems Engineering Requirements

System engineering requirements are categorized in several ways. The following are common categorizations of requirements related to technical management. Operational requirements for systems engineering define the basic needs for answers to the questions posed in Figure 4.1.

Operational distribution or deployment: Where will the system engineering be used?
Mission profile: How will the system engineering accomplish its mission objective?
Performance: What are the critical system engineering parameters to accomplish?
Utilization environments: How are the various system components to be used?
Effectiveness requirements: How effective or efficient must the system be in performing its activities?
Operational life cycle: How long will system engineering be in use by the user?
Environment: What environments will the system be expected to operate in effective activities?

Figure 4.1 Operational requirements.

4.1 Functional and Performance Requirements

Functional requirements include necessary tasks that must be accomplished, actions that needs to be taken, and activities that needs to be performed. Functional review is a process of making decision on what has to be done to fulfill the requirements identified through requirements analysis of a system.

Performance requirements show the extent to which a mission or function must be executed and measured in terms of quality coverage, timelines, and/or readiness. During requirements analysis, performance requirements will be interactively developed to all identified functions based on system life cycle factors and will be characterized in terms of the degree of certainty for system success. System requirements for functional and performance analyses involve defining software engineering needs and objectives in the context of planned use, environments, and identified system characteristics to determine requirements for system functions. Prior analysis will be reviewed and updated along with environment definitions to support system definition. Requirements analysis is conducted iteratively with functional analysis to optimize system's performance requirements for identified functions and to verify that synthesized solutions can satisfy program and project requirements.

The purpose of a system's requirements analysis is to

- Refine objectives and requirements.
- Define initial performance objectives and refine them into system requirements.
- Identify and define constraints that limit solutions.
- Define functional and performance requirements based on provided measures of effectiveness.

The functional review focuses on what the system engineering must do to perform the required operational activities. It includes required inputs, outputs, and transformation rules. The functional requirements along with the physical requirements show the primary sources of requirements that will eventually be reflected in the system specification. The functional review includes information about the following:

- System engineering functions and activities.
- System engineering performance during evaluations.
- System engineering tasks to be accomplished and/or actions to be taken.

4.2 Integrated Team Development

Integrated team development is making a team expertise in operationally employing a product or a software that is being developed. Software developers need to be competent in the operational aspects of the system engineering under development. Typically, systems engineers need to be completely expressed in a way directly usable by software developers. It is unlikely that software developers will receive a well-defined problem from which they can develop system specifications. Hence, integrated team development is a necessary part of a systems engineering team to understand the problems and analyze the needs. The integrated systems engineering team development for functional and requirements analyses is a process of reviews and resolutions.

The following are typical questions that can initiate the thought processes:

- What are the reasons for system engineering development?
- What are the software engineering expectations?
- Who are the users and how do they intend to use products?
- With what environment must the system engineering comply?
- What are existing and required planned for interfaces?
- What functions will the system engineering perform?
- What are the constraints for software?

4.3 Benefits of System Engineering Frameworks

System engineering frameworks can help team members gain more effective plans and procedures to support software engineering services and provide better communications. There should be more creative thinking that enables new perspectives and ways of working, and there should also be a continuous improvement during software engineering design, programming, and development.

System engineering framework makes continuous efforts to define, measure, analyze, improve, and control the aspects that are necessary to achieve stable and predictable results successfully. Achieving software engineering improvement requires commitment, particularly from system engineering support and top-level management. As a result of this framework, it is incumbent on top-level management to prioritize areas of improvement and success.

4.4 System or Subsystem Requirements

Each of the system and subsystem requirements should be assigned a project-unique identifier to support testing and enhance traceability, and they should also be stated in such a way that an objective verification and validation can be defined for it. Project-unique identifiers should use the Program and Project work Breakdown Structure per Contract Work Breakdown Structure (CWBS). Each requirement should be annotated with associated qualification methods. Subsystem requirements should be traceable to systems requirements.

The degree of detail to be provided is guided by the following rule:

- Include characteristics of the system or subsystem that are conditions for acceptance.
- Defer to software engineering design as an acquirer to leave up to the developer.

A systems or subsystem is required to operate distinctly from other software applications. Examples of requirements are idle, ready, active, training, and backup. Itemized requirements associated with each system or subsystem function are defined as a group of related requirements (e.g., subsystem requirements).

Subsystem level includes traceability to the system specification and down to applicable line replaceable units including software Operational Flight Programs or its equivalents. Use of automated tools is highly encouraged, and tools that maintain detailed artifacts of each requirement are preferred.

Summary

Systems engineering framework will be successful when everything is looked at and evaluated to meet compliance for all software development along with management and team employee activities. This framework must always be used to enforce design and programming for companies and institutions, and to ensure that support is implemented, and all tasks and activities are meeting expectations.

Further Reading

Supplementary Text 2001, Prepared by the Defense Acquisition University Press, Fort Belvoir, Virginia 22060-5565. Published by John Wiley & Sons, Inc. 111 River St. Hoboken, NJ 07030-5774. www.wiley.com. Copyright © 2019 by John Wiley & Sons, Inc., Hoboken, NJ. Input Document References; MIL-HDBK-520, Systems Requirements Document Guidance, 5 March 2010.

Chapter 5

Software Engineering Problems and Solutions

To solve or find solutions to problems in software engineering programming, you must make changes to some factors that make these problems being unresolved.

The problems and solutions typically involve complex issues that are difficult to address. Making a change can provide solutions as a result so as to create desired effects on problems. The effects can be embedded in larger structures. The actions taken produce the desired results and needed effects on the problems that need to be fixed.

It is common for a newly developed software to have some problems in the early stages of its development process. Schedule constraints often lead software engineers to adopt and use the "Agile" methodologies that help make the software run correctly and fast.

The ultimate goal is to receive the best response times when the testing teams conduct integration tests.

5.1 Solve Problems

To solve problems, you must be concerned with the following kinds of activities:

- Identifying required results you wish to create and the places where you will measure the extent to which the required results have been achieved.
- Identifying the structure in which these problems and effects associated with points of impact embedded.
- Configuring planned courses of action regarding tasks to be completed, changes to be made, results to be monitored, and adjustments as required.

The solution engineering process is illustrated in Figure 5.1.

Software solutions engineers are network engineers who provide solutions to a variety of problems that occur with software or hardware. When software engineers have to accomplish a task but their network technology is not sufficient to do that, the software solutions engineers come up with solutions to fix it.

Software solutions engineers are a specific type of engineers found exclusively in software companies and institutions. They could find solutions to any problems from computer network installations to customized software programming and development.

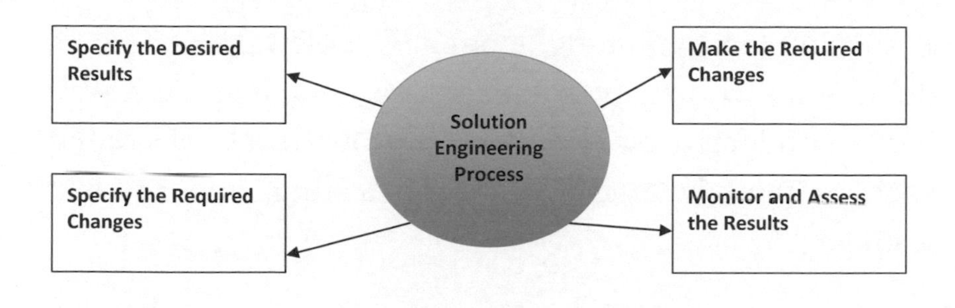

Figure 5.1 Solution engineering process.

5.2 Software Engineering Accomplishments

Software engineering accomplishments begin when a user requests to initiate a specific task that to find solutions for fixing their problems. A software development team segregates user requirements, system requirements, and functional requirements.

Requirements are identified by conducting interviews and based on a database created by studying the existing system. After gathering information about requirements, the software development team analyzes if a software can be made to fulfill all the requirements. System analysis also helps understand limitations of the software product. As per the requirements and the results of the analysis, a software design is made. Implementation of the software design starts in creating software codes in a suitable programming language. Software testing is done while coding by the software engineers, and testing is conducted by experts at various levels of coding such as module testing, program testing, and product testing.

5.3 Software Engineering Problems and Solutions

Teaching software engineering teams about software engineering problems and solutions is a challenging task. Some problems are encountered during teaching the course of software engineering to computer science students and software engineering employees for fixing with appropriate solutions. Software engineering problems are encountered and related to contents of suggested solutions.

The definition of software engineering was first introduced in 1968 at a NATO conference to address a software crisis that came to surface when many large software projects faced great difficulties such as unexpected delay in deliveries. Some of the problems and encountered during software engineering are related to solutions. The definition of software engineering was chosen as being provocative in implying the need for finding solutions to software engineering problems based on practical disciplines that are traditional in the established branches of software engineering.

Other problems encountered are related to many software engineering activities which would lead to the conclusion that most of them have the following contents in common:

- Introduction to software engineering practices.
- Understanding the software development process and life cycle.
- Requirements specifications are important for resolution.
- Implementations, testing, maintenance, and reliability will solve issues.

Software engineering problems and solutions encountered analysis and design structured for objective approaches. This is due to the similarity between some of the software engineering tools used in the programming, design, and object-oriented approaches.

- Package and deployment diagrams.
- Activity and communication diagrams.
- Interaction overview and timing diagrams.

The primary purpose for implementing effective methods in software engineering is to increase communication, knowledge, and visibility in the software life cycle of integration operations. I hope that all readers find this chapter informative, interesting, and convey the information about the methods for software engineering to be more effective in current and future programs and/or projects. The software engineering industry/companies could benefit from adopting these effective methods as well.

In order to develop, operate, and maintain integration capabilities inside software engineering facilities and work products, there must be a major discipline needed for supporting the entire software life cycle to completely understand planning, requirements, design, builds, installations, integration, subcontractors, quality, and delivery. The critical understanding and the start of the right disciplines of these software engineering methods will empower and achieve effective, flexible, and quality results in an integration environment.

Software engineering design, code and unit test, plans, and test procedures integrated with applied software tell us that the software is correctly developed. The software architecture definition provides a framework for creating a software product design and also provides effective methods. The software engineering programming and design also defines implementation details about the software product architecture, components, and interfaces. Element traceability of the software requirements is used by software engineering designers. The traceability data and software engineering programming and design definitions are documented according to program and project plans, ideas, processes, and work instructions.

Software metrics is fundamental to software engineering activities as a discipline. With software measurement, a system is assessed using a range of metrics, and from these measurements, a value for fixes and delivery is obtained. Software metrics can be used to control software engineering process or to predict product attributes. These metrics can control the factors that affect software product qualities as follows:

- Quality of activities related to the production of software engineering tasks.
- Results of the software development, deliverables, and products.
- Measurement of metrics for software engineering performance.

Software programming and design tools are listed in Table 5.1.

Table 5.1 Description of Different Software Programming and Design Tools

Software Programming and Design Tools	*Description*
Requirements analysis and programming and design tools	These tools will be used in organizations by software engineering development programs and projects for requirements analysis of new software. Organizations that do not use the program-wide standards provide documents for inclusion into the program database. Software programming and design documentation retain the tools' format.
Code development tools	These tools have been proved for their effectiveness in designing and developing product line or work product software. The tools such as code editors and compilers are employed.

(*Continued*)

Table 5.1 (*Continued*) Description of Different Software Programming and Design Tools

Software Programming and Design Tools	*Description*
Configuration management (CM) tools	CM tools, ClearCase and ClearQuest, support distribution for incremental development processes implemented in software companies and for military and aerospace program and projects.
Commercial off-the-Shelf (COTS) tools	COTS tools include standard word-processing and graphic development tools to provide for the development and maintenance of documentation with the delivered software.

5.4 Software Engineering Fixes

Software engineer fixes usually require software engineering team members and employees be able to understand codes and ensure that the programming is done correctly. If there is a problem, they write and create a solution using any programming language they feel comfortable with to fix it.

As software engineers, you can fix and solve the following five problems:

- Write software engineering functions that compute the numbers that can be fixed.
- Write a software engineering function that combines lists of elements and document.
- Write a software engineering function that computes the list of problems to be fixed.

- Write a software engineering function that gives a list of non-negative items that form the largest possible number to be fixed.
- Write a function for a program that outputs all possibilities to that the results are always fixed and ready for release.

This chapter helps to understand concerns about software engineering development to solve and fix many software engineering problems. Always challenge the opportunity to solve and fix the problems.

Further Reading

A Brief History of Software Engineering. Online resource www.comphist.org/computing_history/newpage13.htm. Retrieved October 17, 2010.

Mahoney, M., 2004. *Finding a History for Software Engineering*, Online resource www.princeton.edu/~hos/Mahoney/articles/finding/finding.html. Retrieved September 15, 2010.

Naur, P., Randell, B., Eds. *Software Engineering.* Report on a Conference held in Garmisch, Oct. 1968, sponsored by NATO.

Parnas, D. L., Software engineering programs are not computer science programs, *IEEE Software*, November/December 1999, Vol 16, No 6, pp. 19–30.

Chapter 6

Rational ClearCase and ClearQuest

Software engineering and software configuration management (SCM) in general show how tools are implemented using Rational ClearCase and ClearQuest. It is important to considering SCM solutions, and project managers and software configuration managers are responsible for deployment of Rational ClearCase and ClearQuest.

This chapter describes the general concept of SCM and explains why management of software assets and life cycle is a good business. An SCM strategy that leads to the use of Rational ClearCase and ClearQuest products is described in the following:

- Provide the details of SCM using a ClearCase with a focus on test environment, network, servers, and clients.
- Introduce ClearQuest and the roles and responsibilities for users and the infrastructure to provide the details for planning and implementing.

Table 6.1 SCM Improvements

Multi-Sites			
ClearCase and ClearQuest planning	Design and implementation	Software life cycle management	ClearCase and ClearQuest products

- Software change management (SCM) using ClearCase and ClearQuest, design considerations for an effective implementation, and builds for Software Engineering.
- Development in multiple locations using ClearCase and ClearQuest Multi-Sites, including detailed procedures for planning and implementation.

There should always be continuous SCM improvements (refer to Table 6.1 for improvements).

6.1 Software Life Cycle Management

The general concept of SCM explains why software assets and software life cycle management (SLM) is always a good business. Always choose the right SCM and software strategy that should always consider Rational ClearCase and ClearQuest.

I will discuss why it is important to choose SCM tools and processes to handle software development.

Examples try to illustrate the importance of software and its management over its entire life cycle. In fact, SLM has become so important for many organizations and projects whether they are developing systems and applications themselves or they are purchasing systems and applications from software vendors.

To manage software assets, we need to choose the right process and the right tools. SCM tools and processes handle software assets as follows:

- Control is the goal of SCM.
- Software configuration and delivery for process management and problem tracking.
- Software version control is making copies of data in order if necessary.
- Software configuration control implies a higher level of abstraction.
- The SCM tools must be able to identify which versions from a set of components to comprise a specific build.

Software engineering deals with the grouping and manipulation of versions of software progress through the software development life cycle. This typically involves SCM approval levels and production control.

Problem tracking entails recording enhancement change requests, defect reports, and correlating these with the resolution of the request. These reports may include a list of sources involved in the change that can create released products containing desired fixes.

If there are many measurements, it may lead to confusion and distort the overall operations being performed according to SLM shown in Table 6.2.

Table 6.2 Software Life Cycle Management

Software Asset Management	*Developing*	*Purchase*	*Manage*
Management for the whole life cycle	Developing systems and applications	Purchase software applications from software vendors	Managing software assets is a good business

6.2 Why ClearCase and ClearQuest

The most important tools the SCM and software engineering teams can own and use are Rational ClearCase and ClearQuest. Better SCM means better business to show you the benefits of using Rational ClearCase and ClearQuest products as your SCM and software engineering tools of choice.

SCM solutions help implement a managed approach to change that also guards against corruption of assets.

Such tools as Rational ClearCase and ClearQuest provide support for team collaboration and software development, making the tools more productive and effective for working solutions and enabling software development teams to capture, control, and securely manage software changes throughout the whole software life cycle as shown in Table 6.3.

Table 6.3 Manage Software Changes

Manage Software Changes	*Inception*	*Transition*	*Maintenance*
Business modeling and requirements	Analysis and design and implementation	Testing and deployment	Level of control is SCM

SCM with Rational ClearCase and ClearQuest solutions addresses key challenges by the following:

- Supporting software engineering team changes must be seamlessly coordinated for software development projects.
- Addressing key compliance issues arising from SCM, quality assurance, and software engineering process requirements.
- Ensuring that SCM per the plans and processes can quickly improve software engineering.

SCM is a key capability in modern software development practices. It allows software engineering teams to carefully trace requirements over the project life cycle and make numerous changes to them if necessary.

Software engineering delivery is a scalable SCM solution that begins with ClearCase allowing a transition to enterprise ClearCase for Multiple Sites. It also integrates with software engineering full development productivity tools, including Rational ClearQuest having a change and defect tracking tool with unified (SCM) activities based on processes for change management.

Rational ClearCase is the entry-level version; hence, it can be deployed only for supporting some basic SCM functions and processes. Every element of ClearCase versions is run on a single server in the software engineering development life cycle, which helps to track changes, maintain history, and enforce the organization's development process through scripted rules. Thus, using ClearCase facilitates software developers to quickly roll back to previous builds, baselines, or configurations.

Unlike other basic SCM tools, ClearCase supports parallel development through automatic branching which enables multiple software engineering developers to design, code, and test software from a common programming code base. It also includes ClearCase's patented technology to automatically merge parallel branches.

ClearCase is the only entry-level SCM tool to support project organizations with an out-of-the-box process for controlling workflow and unified SCM, and capabilities of ClearCase are shown in Table 6.4.

ClearCase and its products are based on the same fundamental architecture. Therefore, when the software

Table 6.4 ClearCase Capabilities

Capability	*ClearCase*	*ClearCase Multi-Site*
Training required	Yes	
Easy to deploy	Yes	
Support parallel development	Yes	Yes
Support SCM	Yes	Yes
Enforce plans, policies, and procedures	Yes	Yes
Protect code integrity	Yes	Yes
Manageable	Yes	Yes
Enforces development process	Yes	Yes
Accelerated software build management		Yes

development projects grow larger in terms of the number of software developers or the amount of code or artifacts, the organization can easily turn to ClearCase without retraining developers, recreating the programming code base, or rewriting rules and policies.

Transition of software engineering organizations from their previous system to ClearCase will not incur increased costs, delays, disruption, and reduced productivity, which they otherwise experience when changing SCM solutions.

The ClearQuest product provides activity-based change and defect tracking. It can manage all types of change requests including defects, enhancements, issues, and documentation changes with a flexible workflow process, which can be tailored to the organization's specific needs and the various phases of the development process. ClearQuest enables easy customization of defect and change requests, processes, user interface, queries, charts, and reports. Together with ClearCase, it provides a complete SCM and software engineering solutions that cover all areas of the SCM scope. ClearQuest also provides you with "design once, deploy anywhere" capabilities that can automatically propagate changes to any client interface (Windows, Linux, UNIX, and Web).

ClearQuest supports the SCM process and provides for a proven change management process. It also supports projects to run successfully regardless of team size, location, or platform. ClearQuest covers the process management and problem-tracking areas of the SCM domain. It can also be used either as a stand-alone SCM and problem-tracking tool or, together with ClearCase, to provide a complete SCM solution that covers all four areas of the SCM domain. The combination of ClearCase and ClearQuest covers all areas of the SCM domain and provides full integration of activity-based development with process management and problem tracking.

6.3 Rational ClearCase and ClearQuest Platform

Rational ClearCase and ClearQuest platform support aims to align as much possible as with vendor support activities. This support may vary from this strategy due to critical customer needs for building and deploying their software engineering applications and the availability of software and security fixes, or for any other reasons. Rational ClearCase and Rational ClearQuest are focused on providing first-class software support for client–user interfaces.

Both ClearCase and ClearQuest are controlled by SCM specification. To determine which version and element should be visible, ClearCase and ClearQuest traverse the software configuration specification line-by-line from top to bottom.

As a software engineer using Rational ClearCase, you can associate a version with one or more change requests at the same time you check in or check out the version. Moreover, you can also submit queries to identify the change requests that are associated with a program and project over a period of time. As a Rational ClearQuest user, you can view the change set for a request.

Further Reading

Wahli, U., Brown, J., Teinonen, M. & Trulsson, L. *Software Configuration Management: A Clear Case for IBM Rational ClearCase and ClearQuest UCM.* IBM, Corp. Retrieved from http://www.redbooks.ibm.com/redbooks/pdfs/sg246399.pdf

Chapter 7

C++ Programming

C++ programming tools are implemented on the UNIX operating systems, and C++ is a flexible programming language that is popular and used for a large number of platforms. The C++ language is favored by many software programmers because it allows faster programs that are robust and portable. When I organized the Software Configuration Management (SCM) team to implement ClearCase and ClearQuest for the F-22 Raptor program, the C++ programming was used by all software engineers.

The very first thing you need to do, before starting out in C++, is to make sure that you have a compiler. A compiler turns programming that you write into executable to ensure that your computer can actually understand what is being documented. If you're starting out on your own, your best bet is to use Code Blocks with MinGW. If you're on Linux, you can use C++; if you're on Mac, you can use XCode; and if you are stuck using an older compiler such as Turbo C++, you'll need to read this chapter for compatibility issues. If you haven't done so, go ahead and get a compiler set up for use.

7.1 Introduction to C++ Language

C++ programming language gained popularity as it was adopted by numerous programmers around the world and some software programmers added it to their language of choice. A committee of the American National Standards Institute (ANSI) and the International Organization for Standardization (ISO) has defined the standard version of C++ as portable to the platform and to developed environments.

The C++ program is a collection of commands that allow the computer to execute the activities for using the C++ software programming. This collection of commands is usually called C++ source code and it includes commands for functions and/or keywords. Keywords are basic building blocks of the language, while functions are usually written in terms of simpler functions you see in first programs. C++ provides a great many common functions and keywords that you can use. But how does C++ programs actually start? Every program in C++ has one function named `main()` that is always called when your programming is first executed.

7.2 Commenting on Programs

To access standard functions that come with the compiler, you have to include a header that will include directions. As you are learning software programming, you should also start to learn how to explain your programs and add comments to software code to use frequently to help explain all code examples. Be certain not to accidentally comment out a piece of code. When you are learning to be a software programmer, it is useful to be able to comment out sections of software code in order to see how the output is affected. Installing the complier will execute C++ programs into byte software code that is understood by the computer.

7.3 Standardization of C++

The standardization of C++ programming was adopted by many software programmers around the world. Programs and organizations run in a console windows on Linux systems to demonstrate the mechanics of the C++ language. A software coding generated by a visual development tool on a Windows platform can make you familiar with the C++ language and be able to edit it to create windowed applications.

7.4 Compiling for Running Programs

When C++ succeeds, the compiler for running programs creates executable files for software source code files. Compiling software source code files in the same directory would overwrite executable files, so a customer's name is specified for each executable file when working in programs using the compiler options.

For programming, you have to do the following:

- Move the program source file to a program directory on your system.
- At command prompts, use the command to navigate to the "My Programs" directory.
- Enter a command to attempt to compile program using C++.
- Also enter a command to compile programs to execute files alongside source files such as C++ hellow.cpp – hello.exe.

To run an executable software program file generated in Windows, you must enter the file name when you see the My Programs directory to have the file extension implemented (see Table 7.1).

7.5 Programming Files

The ability to software programming files provides useful methods for maintaining data on computers. The standard C++ software programming provides functions for working

Table 7.1 Command Prompt

C:\MyPrograms> C++ hello.cpp – o hello.exe
C:\Users\Name> cd C:/My Programs
C:\MyPrograms>hello – Hello World!
C:\MyPrograms >_

files which can be made available by directing at a start of programming. For each file, a file stream object must be created for writing data to the programming file for help in reading data from the file. The file stream object for programming begins with of stream keyword followed by the name of the programming file to be written to.

Before writing to the output to a file, the program should test to ensure files have been created and performed to allow the programming to write output when the test is successful. While starting a new program using C++, you should start library classes to do the following:

- Add main functions containing statements for software programming.
- Create output file streaming objects for software programming.
- Insert statements to programming files to compile.

7.6 C++ Programming Elements

The time function is the number elapsed, and it ensures that numbers generated will appear to be true for random use. To start a new program, you should do the following:

- Specify library classes to include random support.
- Add the main function containing return statements.

- Keep the random number generator for current time use.
- Save, compile, and run the program and project to be successful.

Summary

C++ is an object-oriented programming language that is an extension of software programming language, and each method must contain a main method as the entry point to all programs and organizations. The C++ commands call the compiler to specify generated executables to store multiple items of data in elements. It is recommended that software programming code be widely commented to make sure the purpose is clear. The visual programming with C++ provides functions to work with graphical controls. Once programming is completed, the program or project should be tested to ensure that the methods involved meet all the requirements of program and project plans. C++ applications can run on any machine and laptop with an appropriate .NET framework.

Further Reading

McGrath, M., 2015, *C++ Programming in Easy Steps*, Warwickshire: In Easy Steps Limited. ISBN 978-1-84078-432-9. www.ineasysteps.com.

Chapter 8

Software Configuration Management

The purpose of Software Configuration Management (SCM) is to provide a common operating framework in which activities such as building software and documenting SCM plans, and Version Description Documents (VDD) can be shared between management, team members, and software quality engineering (SQE) representatives. The expected SCM building of software is delivered to test labs for verification and validation by SQE. Components of SCM is making sure software configuration has identification, change control, SCM status accounting and SCM audits and reviews are performed.

For 20 years, I was an SCM engineer working for the Boeing Defense and Space programs for B-2 Stealth Bomber, F-22 Raptor, Advanced Systems, and Airborne Early Warning & Control (AEW&C) in Australia. I was able to put together an SCM team to get ClearCase and ClearQuest installed when working for AEW&C and F-22 Raptor programs.

8.1 Software Configuration Management Items

Software configuration management (SCM) items are the software while under the development for programs, data, documents such as the software requirements specification, test cases that are referred to as SCM items during the software engineering development.

SCM provides for systematic evolution of a software under development and provides for:

- Visibility
- Controlled change
- Traceability
- Monitoring.

8.2 Goals for Software Configuration Management

The goals for an SCM must include planned activities to ensure that software work products are identified, controlled, and made available. Changes to identified software work products are controlled, and affected groups and individuals must be informed of the status and content of all software baselines. Baselines include the following:

- Phase and discipline
- Requirements analysis

- Software design and specification
- Software programming and implementation
- Testing and integration
- Test plans and data
- Release of operational software.

The baseline is a specification or a product that has been formally reviewed and agreed upon and serves as the basis for further software development.

SCM goals are provided in Figure 8.1.

SCM identification includes setting and maintaining software engineering baselines, which define the architecture, components, and software developments at any point in time. It is the basis by which changes to any part of SCM are identified, documented, and later tracked through software programming, design, development, testing, and final delivery to labs and customers. SCM incrementally establishes and maintains the definitive current basis and

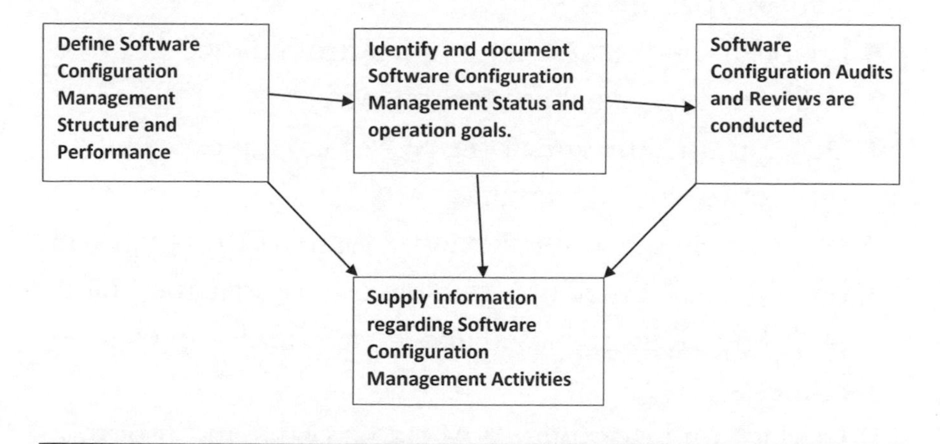

Figure 8.1 Software Configuration Management goals.

its configuration items (CIs) throughout their product life cycle from software engineering development, production, deployment, to operational support. Software configuration control includes evaluation of all change requests and change proposals, and their approval or disapproval. SCM covers all the processes that control any modifications to the designs of hardware, firmware, software, and documentation.

The information about the effectiveness of a software engineering process is necessary for management, team members, and employees, and so it should be made available through SCM engineering. Software changes should be documented, and the information should always be checked for any updates. Updated information for the CIs is available in the SCM database.

The responsibilities for SCM are as follows:

- Implement the SCM program and SCM requirement for software baselines.
- Establish and maintain a software change request (SCR) tracking database for reviews.
- Develop and document the SCM plans and operating procedures and documents.
- Form an Engineering Review Board (ERB) that will prepare and distribute its agendas, record the status of SCM change requests, and review and approve the requests.
- Produce and distribute SCM status plans and reports.

- Coordinate SCRs with respective individuals to implement approved changes and to resolve and fix all SCR problems.
- Make sure that software quality representatives (SQR) are included in all ERB meetings.

8.3 Integration Build and Release

The role of an integration build and release engineer is to construct software systems and set up software components. They combine the work of many SCM teams and assemble software builds for integration and release. Software integration is the process of bringing the works of different SCM teams together, testing them, and making adequate changes to produce software systems with required performance.

There are two types of integration used by SCM: merge and assembly integrations. Merge integration involves the solutions of changes made by SCM teams to common files and components. It should always be remember that merge integration requires the knowledge of changes made to software systems.

The integration build and release methods provide required steps to be conducted for integration and checkout of informal software engineering builds. The SCM engineers, software design/development teams, and test engineers need to develop a strategy for planning, design, execution, data collection, and test evaluations.

The software builds and integration activities are formal and flexible and prepare for the software and systems integration phase for the work product.

The strategy for software integration builds and release provides a road map that describes the steps to be conducted as part of the implementation of software to start integration activities. When a strategy is planned, adequate resources are required. This strategy should be flexible and promote an approach that could show change.

Sometimes, planning by a senior engineer, program and project managers need to track the progress of program and project, and also need to do the following:

- Conduct effective technical reviews.
- Show different integration techniques and software approach.
- Involve software engineering designers from the start to the end of the work in SCM.

8.4 Assembly Integration

Assembly integration provides baselines for software components that could be large pieces of the overall software system. Assembly integration can occur during software build times and SCM activities, and can bring together source components of software builds. Different types and levels of assembly integration are used based on the size of the software build systems. At certain levels, software

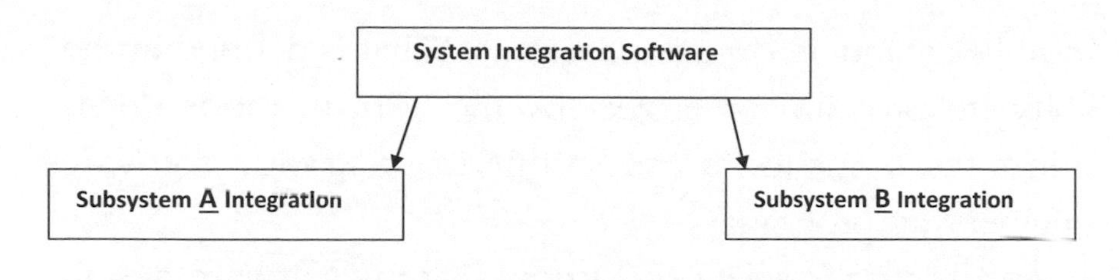

Figure 8.2 Software system integration.

systems must have a well-defined architecture to use the assembly integration.

There are three assembly integration scenarios required for the SCM teams to develop the software system independently and perform assembly integration.

The first level of assembly integration is referred to subsystem, and the second level is software system integration for A and B as shown in Figure 8.2.

8.5 Software Configuration Management Methods

SCM methods are used for the development of software engineering programs and maintenance of computer software engineering tools. There are a variety of SCM methods used by many software programs worldwide. The terminology, definitions, or terms used by SCM can be different depending on what software program you work in. An example is; ERB and SCR can be used and defined. Another definition is Version Description Document (VDD), which I have seen and used in all my 31 years of working in military and aerospace programs. What is defined in the software program or project by the SCM methods determines the terms to be used while working on a software engineering program.

As a software designer, I know that my software design activities are important and comfortable to release software code for SCM to configure and control along with building software code changes or updates to release for test, verification, validation, and delivery to customers. There are more about SCM organizations that software designers do not know. Many software designers feel SCM has no understanding of software engineering, and I beg to differ that there are numerous software designers who want to know the role of SCM.

The SCM activities are necessary to

- Establish and maintain the software engineering identification process and control changes to identified software products and their related integration and documentation.
- Record and report information needed to manage software products effectively, including the status of proposed changes and the implementation status of approved changes.
- Maintain auditable records of all applicable software products help verify conformance to specifications, interface control documents, contract requirements, and as-built software configurations.

Summary

The SCM build engineer will perform tasks related to software construction and configuration control that include creation of a folder to store documentation of the build, software source code changes, and records of computer program development. Software build requests (SBRs) are written, and a build procedure checklist is provided to assemble, compile, and link source codes; build archive copies; provide listings for use in software engineering development, test, and delivery support; and document the VDD.

Further Reading

Berczuk, S. P., Appleton, B., 2003, *Software Configuration Management Patterns: Effective Team-Work Practical Integration*, First Edition. Boston, MA: Addison-Wesley. ISBN 0-201-74117-2.

White, B. A., 2000, *Software Configuration Management Strategies and Rational ClearCase: A Practical Introduction*, Boston, MA: Addison-Wesley. ISBN 0-201-60478-7.

Chapter 9

Software Quality Engineering

All companies, institutions, military, and successful businesses need to advance and improve the performance of software quality engineering (SQE) to ensure that builds, changes, and weekly meetings of software configuration management (SCM) are compliant with all software needs. A critical part of improving SQE performance is enabling software management, SCM team members, and software engineers to report issues and events that might affect their ability to deliver software with high standards.

The SQE Performance and Improvement Reports are the basis of quality efforts to look constructively at how can software services and workplace be made safer, more secure, more efficient, and more consistent. The purpose of these reports are not to find fault with or blame at anyone; rather, it helps to be more mindful about what is done by software engineers and SCM and how to accomplish and identify aspects that might need extra support to achieve the highest levels of software engineering deliveries.

9.1 Software Quality Engineering Progress

There are many stages in progress in which software quality engineers would guide software engineers and SCM to know and play their roles and responsibilities to ensure that the performance of SQE and SCM has improved to provide better outcomes. In my software engineering experience of 31 years, being better and successful and ensuring that processes of performance and improvements are compliant with requirements have always been important.

Software engineering audits, reviews, and compliance requests are common when program and project core software applications are up for renewal. As a starting point in processes, software engineers will require to prove that they are licensed to play their respective roles and responsibilities. Most software engineering terms and conditions require to voluntarily participate in any software audits, reviews, or compliance requests.

9.2 Preparation for Software Quality Engineering Performance

There are guidelines and values for preparation for SQE performance based on the performance and ability of software engineers to prepare daily ideas and concepts that help accomplish the software engineering tasks.

The main ideas relate to making correct statements to identify and rectify software engineering issues and ensuring schedules are met for companies, institutions, military, and successful businesses. The following are few examples:

- Capability for performing software engineering is required.
- Important schedules must be maintained and effective with management/SCM/SQE support.
- A software quality engineer performance is a type of measurement.
- Performance evaluates the success of software, programs, projects, products, plans, etc.
- Choosing the right software engineering performance relies upon a good understanding.

There is a need to understand what is important for various software engineering techniques that are assessed to their key activities. The SQE assessments often help identify initiatives for potential improvement. A very common way to choose performance is to apply a SQE framework.

The SQE framework will establish standards relating to software engineering performance improvement with respect to facilities and provide technical assistance in order to meet such standards. Best practices include how to coordinate the implementation of plans and procedures with software quality assessment and assurance activities.

9.3 Preparation for Software Quality Engineering Improvement

Preparation for SQE improvement ameliorates capabilities and key results of management and software engineers for having successful objectives and achieving better results. SQE plans provide a sequence of actions to be implemented in order to improve work process of companies, institutions, military, and successful businesses to reach successful accomplishments. Management needs to establish a proper channel of communication to lead and show software engineers involved in decision-making their commitment to carrying out all decisions. The SQE improvement is a simple process that involves basic common sense and a simple logical approach to improve principles and techniques that help fulfill the responsibilities of software engineering.

In order to provide audits, reviews, and evaluations for Software Quality Assurance Engineering processes look at the time spent in each software engineering development phases. When management, software engineers, and SQE personnel spend detailed time, it shows that evaluations are reasonably complete. To achieve high SQE tasks, always spend time to ensure that review times and inspections measure the importance of performance and improvement.

I suggest using the following guidelines:

- Use contract and requirement reviews for inspections.
- Conduct high-level SQE audits, reviews, and inspections.
- Ensure data audits and reviews for release are compliant.

It is effective to use SQE checklists and update them frequently with issues or concerns you see related to discrepancies in software engineering and SCM. When data reflects problems always review deep content to find the cause. After the completion of SQE evaluations, they will show where adjustments need to be made. Strive to improve process and SQE to be successful during all activities performed by software engineers and SCM.

9.4 Preparation for Quality Assurance Improvement

Roles and missions describe the scope with preparation for quality assurance improvement capabilities. Key results for QA improvement relate to management and employees for having successful objectives and for achieving better results. Action plans provide a sequence of actions to be implemented in order to improve work process of companies, institutions, military, and successful businesses to reach successful accomplishments. Management needs to properly communicate with software engineers involved in decision-making regarding their commitment to carrying out all decisions. Quality assurance improvement can be a simple process that involves basic common sense and a simple logical approach that improves principles and techniques to fulfill the responsibilities of software engineering.

9.5 Software Quality Engineering Excellence

Management and software team members can help in the drive for SQE excellence and always show the importance of quality for all companies, institutions, military, and successful businesses. Make discipline a habit to ensure that knowledge and skills achieve all expectations required by management software team members.

To achieve excellence, the following points should be taken into account:

- The knowledge of software and design development is important.
- Software quality engineers must have the skills that they learned through lessons to accomplish their responsibilities.

The desire for striving for excellence should be motivated to be consistent in all companies, institutions, military, and successful businesses.

Note: When a manager or software engineers do not follow processes, there should be a pressure applied to make sure that conformance is known and is important. Motivation and maintaining disciplines help establish those processes, plans, and goals and monitor these actions on a daily basis, and a pressure applied can be effective for achieving excellence.

Achieving excellence can be a struggle, since the world keeps changing and facts are astounding each day. Human capabilities increase every year, and the key element is that humans excel. When people work to excel, they often find that they can work better, strive to improve, and have better management and employees responsible for handling problems. Software quality engineers can help management and employees to work things out and to step up to all challenges and convincing SQE plans and have management agree to what is proposed. Software engineers must care about the quality of all products designed and developed, and be consistent with all effective methods of software engineering to reach that capability. Excellence can start from software engineers because knowing that the software engineering programming is changing and the focus on improvement is important is necessary to be successful.

There should always be a model for defining software quality process with a step-by-step direction toward effective improvement. The process for SQE improvement stages and levels can be achieved by using an example such as the Capability Maturity Model Integration (CMMI). The results of assessments should always identify the strengths and weaknesses of software quality engineers for improving processes to be effective and compliant. There are few symptoms of having effective process improvement such as

- Discipline will ensure management and team members follow processes.

- Be effective and make sure conformance is checked and issues are reported.
- Visibility of software engineering process documentation should be effective.
- Evaluations of software engineering performance are linked to process goals.

9.6 Software Quality Engineering Problem-Solving

SQE problem-solving occurs when the team ignores peer pressure from management on aspects that can cause threats and discouragement. While addressing these SQE problems, the team should review these aspects and be positive to help resolve the concerns of the management. If problems are discussed and responded quickly, it is important for the team to call for help to resolve difficult problems. Never feel that you are alone and insist you will fix problems with others. Support from other team members is critical because it will provide you time in helping management and employees resolve problems provided to you and your team members. This is key to provide effective support and show that team members believe in their abilities to solve quality assurance problems and do superior work. Remember that team members will always believe in themselves and will often perform beyond what can be very possible for solving all problems, issues, and concerns.

9.7 Common Software Quality Engineering Working Framework

Having a common SQE working framework is important where SQE team goals could be challenged and the path to achieving goals must always be clear. All tasks need to be achievable and the roles and responsibilities must be understood to accomplish the team goals. The tasks needed for common SQE working frameworks are as follows:

- Software tasks to be performed and accomplished.
- Provide a timeline for all software tasks to be completed.
- Assign software engineering team members required tasks to achieve completion.

All software engineering teams need to be cohesive and have challenging goals, performance, feedback, and agreements for the framework processes to be defined for all activities for management and employee accomplishments.

The SQE working framework is intended to improve the software quality of general practices and be part of an effort to solve issues and numerous problems. Participation is required for most activities under the contracts and it can make a difference in daily processes implemented for SQE. The criteria for SQE framework activities are designed around best practice and have a number of accomplishments allocated for achievement.

In many companies, institutions, military, and successful businesses, the value of accomplishments is always measured for expected goals. The working framework is reviewed, audited, and evaluated by SQE representatives to ensure that processes, plans, and procedures are working to achieve high expectations.

9.8 Software Quality Engineering Impacts Business

It is sometimes easy to ensure SQE impacts business when working and meeting program and project deadlines. SQE representatives can support and understand turbulent outcomes than can arise if required standards are not enforced. It is clear that businesses fail when companies, institutions, military, and business operations fail and when policies are not followed. Management, team members, and employees have opportunities to turn situations around that requires a strong understanding to improve and implement changes and have open communication for problem-solving.

Here are five actions that can be used:

- Develop leadership to champion SQE practices.
- Ensure that everyone knows the guidelines and expectations.
- Communication is important as a top priority.
- Gain feedback from employees who bring up concerns.
- It is essential to provide a strong understanding of processes required.

9.9 Ways to Improve Software Quality Engineering

It takes courage to make clear objectives for ways to improve SQE. It's worth to create an effective understanding of competing interests for management and software team members and to maintain focus on programs and projects. Always help everyone to stay on track using these six tips to move faster and more efficiently:

- *Set a Software Schedule* – Review task lists and have a huge list of ideas and solutions that can be acted for positive actions.
- *Invite Team Members' Help* – Enter SQE tasks into programs and projects to make sure every step is accounted in order to limit the possibility for software programming and design.
- *Make Your Meetings Count* – All meetings are important to be accomplished and include updates on software engineering programs and project management systems on a regular basis.
- *Communicate Clearly* – It is important to understand requirements, expectations, and tasks for software engineering constantly coming to SQE with more questions.
- *Focus* – If you want to move your programs and projects forward, focus on both at a time and be flexible if you have dependencies that are waiting to get finished.

- *Goals* – All software engineering goals need to be accomplished and all programs and projects performed to compliant.

Summary

This chapter provides an overview of what SQE teams need to do to resolve problems. When software tasks and activities conducted fail, it is usually due to software and technical problems. The criteria for successful SQE team tasks must be clearly understood, I organized a high-performance work team (HPWT) that drives success through audits, reviews, evaluations, and solutions to ensure complex software engineering tasks are correct and current for the competitive work environment activities and to be ready

Table 9.1 Definitions of Software Quality

Software Quality Assessment – is an evaluation of a process and plans to determine if a defined standard of software quality is being achieved.
Software Quality Assurance – is the organization's structure, processes, plans, and procedures designed to ensure that all software practices are consistently applied.
Software Quality Improvement and Process Performance Improvement – is an ongoing interdisciplinary that is designed to improve the delivery of services and outcomes.
Software Quality Process – empower management and software team members to develop solutions per software development and design.

for delivering software engineering products. Management supported me putting the HPWT together.

The pleasure of being successful sustains the enthusiasm and the importance to have management, team members, and employees to work processes and ensure all processes followed, as shown in Table 9.1, show definitions of software quality.

Further Reading

Broughton, R., 2008, Quality assurance solutions, USA, online information. https://www.academia.edu/5642662/Broughton_et_al._2008.

Chemuturi, M., 2010, *Software Quality Assurance: Best Practices, Tools and Techniques for Software Developers*, Plantation, FL: J. Ross Publishing. ISBN 978-1-60427-032-7.

Doherty, G., 2012, *Quality Assurance in Education*, London: IntechOpen. doi:10.5772/32434.

Endean, M., Bai, B., Du, R., Quality standards in online distance education, *International Journal of Continuing Education and Lifelong Learning*, 2010, Vol 3, No 1, pp. 53–73.

Chapter 10

Verification and Validation

Verification and validation are independent procedures and processes that are used together for checking the quality of software engineering products. Software configuration management (SCM) builds, services, and systems engineering must meet requirements and specifications that fulfill intended purposes. The words "verification" and "validation" are prefixed by "independent" indicating that verification and validation are to be performed by software quality engineering (SQE). "Independent verification and validation" can be abbreviated as "IV&V."

10.1 Verification

Verification is intended to check that a software engineering product meets a set of software design specifications. In the software development phase, verification procedures involve special tests to model or simulate software products performing a review or analysis of the modeling and buy-off results.

Verification plans and procedures involve regular repeating tests devised specifically to ensure that the software engineering product, service, or system continues to meet the initial software design requirements, specifications, and regulations. It is a process that is used to evaluate whether a product, service, or system complies with regulations and conditions imposed at the start of a software engineering development phase. Verification can be conducted in stages of software development, scale-up, or production as an internal process.

Verification of software and equipment usually consists of design qualification (DQ), installation qualification (IQ), operational qualification (OQ), and performance qualification (PQ).

Verification is performed by an SQE by confirming through reviews and testing that the software and equipment meets the written acquisition specification and requirements. The relevant plans, documents, and manuals of software/equipment are provided to be thoroughly performed by the software users who work in regulatory environments.

10.2 Validation

Validation is intended to ensure a software product, service, or system results meet the operational needs and expectations. Validation plans and procedures involve modeling using simulations to predict faults or gaps that might lead to invalid or incomplete verification or development of a software product, service, or system.

Validation requirements, specifications, and regulations are used as a basis for qualifying a software development flow or verification flow for a software product, service, or system. Additional validation plans and procedures are also designed specifically to ensure that modifications made to an existing qualified development flow or verification flow will have the effect of producing a software product, service, or system that meets the initial design requirements, specifications, and regulations. Validation helps to keep the flow qualified and its processes establishing evidence that provides a high degree of quality assurance that a product, service, or system accomplishes its intended requirements. This often involves acceptance of fitness for purpose with software engineers and other product stakeholders. It is entirely possible that a software product passes when verified, but fails when validated. This can happen when a product is built as per the specifications, but the specifications themselves fail to address the software engineering needs.

10.3 Verification and Validation Test Plan

The verification and validation test plan describes how software engineering products will be tested and defines the items that are to be tested. These items may be the requirements documentation or software engineering design documentation. There could be different names for these documents, but no matter your testing comes out of these documents. You have to figure out not only what tests you need to run but also what procedures need to be in place for the test. A test has planned inputs, so think about what the expected results are based on those planned inputs.

Software testing consists of various tasks, such as defining testing schedules and creating test cases that need to be completed in order to carry out the compliant testing. Elements of the testing tasks are testing plans, test design specifications, test cases, test procedures, test results, and defects. The details in a specific environment in which the test is conducted will show how the solution works in the environment. Look into the nonfunctional requirements to determine test needs such as software test, configuration test, and so on.

Knowing where the tests are conducted is important to developing the test plan. You want to consider the following:

- If the test is in a lab, know the lab conditions.
- Check environmental concerns that could affect the test.
- Test areas must be accessible to the testers for verification and validation.

10.4 Software Engineering Verification and Validation

Software engineering verification and validation is the process of investigating whether a software system satisfies specifications and standards and it fulfills the required expectations. The scopes of verification and validation are as follows:

- *Verification*: Are we building the product right?
- *Validation*: Are we building the right product?

Software engineering verification is the process of checking whether a software engineering achieves its goal without any issues and concerns. This ensures products developed are right. It verifies whether the developed product fulfills the requirements and expectations.

Verification is static testing of activities involved in verification through

- Inspections.
- Reviews.
- Walkthroughs.
- Desk checking.

Software engineering validation is the process of checking whether the software engineering product is up to the performance or in other words has high-level requirements. It is the process of checking whether software engineering products are developed in the right way, and validations of actual and expected products are carried out by Dynamic Testing. Activities involved in validation are

- Black box testing.
- White box testing.

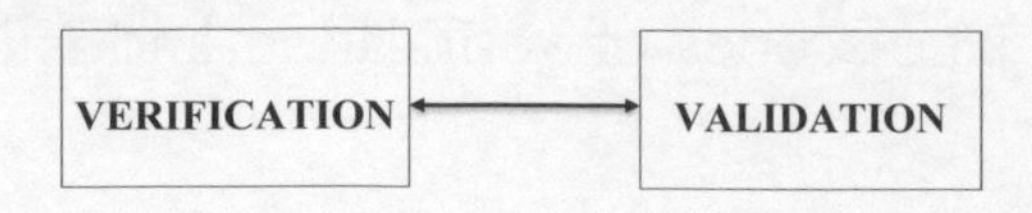

Figure 10.1 Verification and validation.

- Unit testing.
- Integration testing.

Verification and validation are shown in Figure 10.1.

Summary

My third book titled *Effective Processes for Quality Assurance* will describe the effective processes and activities performed in support of verification and validation. The purpose of verification is to ensure that software products meet its specified requirements, which is accomplished through various activities ranging from in-process reviews and audits. The purpose of validation is to ensure the work product fulfills its intended use when placed in its intended environment. SQE representatives will audit all activities and support qualification by monitoring all activities. The following tasks and activities must be performed:

- Analyze data from verification and validation to show expected results.
- Ensure correct standards of products submitted for verification and validation.

- Document all results of each test activity.
- Coordinate verification and validation results with software team members.

After verification and validation, reports are developed and written, and requirements are determined, prepare compliance for closure to plans, procedures, and documentation to show proof of compliance and to be ready for release.

Further Reading

Difference between "Verification and Validation. *Software Testing Class*. 2018. In interviews most of the interviewers are asking questions on 'What is Difference between Verification and Validation?' Lots of people use verification and validation interchangeably, but both have different meanings. John Wiley & Sons, Inv. 2019.

McCaffrey, J. D., 2006, *Validation vs Verification*. https://jamesmccaffrey.wordpress.com/2006/04/28/validation-vs-verification/.

Chapter 11

Management Support for Software Engineering

11.1 Managers Are Important

Having great managers for providing management support for software engineering is important to achieve success and maintain a good software engineering team. In general, a great manager is seen as enabling the software engineering team they manage to show their full potential. Software engineering research studies factors that may affect the performance and productivity of software engineers using tools and skills overlooked by software engineering managers.

Experts are questioning how the abundant work in management applies to software engineering. On the other hand, practitioners are looking to researchers for evidence-based guidance on how to manage software engineering teams. Many researchers conducted mixed methods empirical studies of software engineering management at many companies to investigate what managers attribute to software engineers and the importance of identified attributes. The conceptual framework of manager attributes and finding technical skills are the sign of greatness for software engineering managers. Through statistical analysis, managers relate their views and how software engineering works in teams to make their management great.

The responsibility of management and software engineers is to deliver products that make the software engineering companies successful. This is generally captured by various metrics of productivity, performance, and success. The managers are also responsible for creating conditions where software engineers can feel motivated and productive. Understanding what impacts software engineering perceptions of their managers is of high importance. The goals are to understand how software engineering managers function and what is perceived to make them great and supportive. Great managers positively impact motivation and engagement, and aim to raise awareness of software engineering aspects, as they can affect software engineering outcomes. There are attributes that are perceived to characterize great software engineering managers and why these attributes are important. After decades of research on software management in general, there are a large number of theories that describe these managers and their behaviors in companies.

11.2 Conceptual Framework for Engineering Managers

There are high-level attributes of great description for each conceptual framework for great software engineering managers in Figure 11.1.

The software engineering manager must interact on two levels with the individual software engineer and with the software engineering team. The vertical, dashed-lined grouping in the framework shows the relevant attributes for each interaction. Conceptual frameworks for software engineering managers are particularly useful as organizing devices for use and research. Software engineers use the conceptual framework to distinguish the behavior and incentive systems for all software case examples of a macro-level conceptual framework.

Managers tend to believe that employees such as software engineers are the most important part of their job. Sometimes, it is necessary for software engineers to consume time and always be available for correcting the real software problems. Performance management for software

Manager Functions	Build relationships with Software Engineers	Guide the Software Engineering Teams
Manager Functions	Support the Software Engineering Teams	Provide Communication
Manager Functions	Maintain Positive working communications	Motivate important solutions and support

Figure 11.1 Conceptual framework for great software engineering managers.

engineering becomes the strongest link in the chain of management activities that keep companies and businesses profitable. Managers typically do receive any special training regarding performance development and understand how to help their software engineers to become more successful. This is why a performance management operation is required to maintain all the management activities.

A manager's performance is a structured approach for improving the effectiveness of all employees and ensuring software engineers are do the job right. Approaches for enhancing job effectiveness are in a structure that emphasizes the use of standardized tools and procedures such as a performance evaluation process that achieves dramatic organizational results in a timely manner.

Providing effective management in an organization and program is like maintaining a wheel that can handle any terrain and can go great distances without breaking down. One of the main activities of management is improving performance through proper observation, assessment, and performance feedback. I call this as "performance development" because its main objective is to manage employee performance and help develop job-related competencies.

11.3 Working with Managers for 31 Years

When I worked for Hercules Missile Defense and Boeing Defense and Space for 31 years, there were software managers, software engineering teams, software configuration management engineers, software quality engineers, systems engineers, lab test engineers, flight line representatives, software suppliers, customers, and military air space representatives working with me. It was great to work with these people, and we made everything work well. The managers were very good and were leaders.

The coordination with software managers went very well, and they were always there for us to make sure that the plans, procedures, and processes were being followed. There were also weekly meetings conducted to ensure that all activities were being done correctly and ready for use.

It was great to work with the managers at Northrop-Grumman for the B-2 Stealth Bomber and Lockheed-Martin for the F-22 Raptor Jet. I was working there with the companies to make sure once in a year that our Boeing Defense and Space software was working correctly in the Air Force Aircrafts.

B-2 Stealth Bomber

F-22 Raptor Jet

Software engineering programmers and developers would face difficulties if their managers don't know anything about software development, which would make software engineering way more complicated.

Managers must think they know what they are doing and how hard is it to just set a deadline and expect software engineers to meet schedules. Software programming and development has a particular way of working that will just make sense to technical managers. Managers who are experienced and have a software engineering background often think software engineers will work faster if they introduce effective software engineering into the projects being worked. A senior software engineer will get the new software engineer up to speed on the team's progress.

There could be problems that can get even worse if that senior engineer is a new hire, and he or she will have to be double-checked by one of the smart engineers to have valuable time spent working the actual software engineering task.

Software engineering programmer requests from managers are even possible and could become a way that does meet everything in the programs and projects. Managers can support software engineers to accept requests and commitments by providing them more time than the software manager originally quoted and increasing the capabilities of the software engineering teams. Software managers who are technical ones seem to think that any software programming that goes through software quality engineers should undergo thorough verification and validation before they are completely ready for delivery. The goal then would be for quality engineers to spend hours each day on software testing. At times it is hard to test complex software programs because there are so many variables to work with and many file types being tested.

Another important aspect of software programming is to behave logically because programming bugs can be triggered under the most specific and outlandish conditions if software quality engineering can't predict every possible behavior and condition. Working with software engineering programmers provides opportunities to interact more with the programs and projects.

Many software managers seem to think that if software engineers are spending more time, then the programs and projects are moving forward. When most software managers are faced with a choice between doing something, the products still work and the software manager looks good for exercising "management skills" to get the programs and products completed.

11.4 To Be a Good Manager

Good managers come in all directions, and it is always easy for them to find appropriate software engineering methods. If managers are interested in pursuing leadership roles, it is important to start by asking themselves why they want to be a manager and leader. Answering questions given by software managers to software engineering team members is not necessarily a bad thing, but wanting to serve others is one of the best signs of an effective leader and manager.

Part of leadership is a courage proposition of defending the software engineering team and representing your team and taking risks. Being a manager is not about doing more of the work or being the most technical person on your team. Rather, it is actually about giving up some of your work to help others grow. It's about being a support system and sticking up for your software engineering team when experiments go sideways. Being a manager is about knowing when to hand off the reins and help new leaders and software engineers flourish.

As a software manager, it is important to have software engineering skills. Software managers can empower their software engineering teams to grow by coaching them. From the skills perspective, software managers should not just tell software engineers how to act, but they should also figure it out themselves. Software leaders often want to share their hard-earned experience with others and always provide the best way to help them learn.

11.5 The Basics of Software Engineering Management

Many companies track their software engineering assets for reason to know what they have and what they need. An effective software engineering management requires comprehensive and uniform practices to support processes and technology to ensure software compliance, manage software costs, improve security, and make better program and project business decisions. A basic function of software engineering management is to provide inputs for managing the total activities around all of an organization's technology assets throughout their software engineering life cycle.

Software engineering management is about capturing a deep and wide view of all software assets across your entire programs and projects. Whenever a software is installed to be used for whatever purposes, make sure who is using the software. The extent to which the software is used helps track and manage its programming across programs and projects. Technology quantifies software engineering assets on an ongoing basis, and it must provide an automated way for software engineering leaders to understand their software user database and software engineering versions in order to gain clarity about assets using Windows, Linux, or Mac environments. In this way, software engineering managers can reconcile system use against software to help with external software engineering audits and reviews.

11.6 Software Engineering Management Connections

Software engineering management connections show strengths and what is needed to be worked on and socialized together regularly. A sense of connection can run very high and be sustained over time. A software engineer can deliver value by creating a strong sense of connection. A good software engineering manager must be a good listener. Good listening skills are especially important in the software engineering testing domain. This is a skill worked on from the beginning for professional careers.

Software engineering managers like to see results from team members. It's important to connect and listen in order not to be influenced too much by information. Software engineering managers must lead by examples of actions than words, which would impact their team positively. This can be done by hands-on software engineering testing and connecting with clients, employees, and team members.

Chapter 12

Agile for Software Engineering

Agile for software engineering is a time-bound, iterative approach to software delivery that builds software incrementally from the start of the program and project, instead of trying to deliver software performed all at once.

12.1 Why Agile?

Nowadays, software engineering technology is progressing faster than ever, enforcing the global software companies and institutions to work in a fast-paced changing environment. Because these programs and businesses are operating in an ever-changing environment, it is important to gather a complete set of software engineering requirements. Without software engineering requirements, it becomes practically hard for any conventional software model to perform software engineering activities.

The importance of software models such as Agile depends on completely specifying the requirements, designing, and testing the system toward rapid software engineering development. As a consequence, a conventional software engineering development model will deliver the required products needed.

Let's now read about the Agile principles laid out for its foundation:

- The highest priority is to satisfy customers through early and continuous delivery of valuable software.
- Must always deliver working software frequently from weeks and months, with a preference to the shortest timescale.
- Build software engineering programs and projects around motivated software engineers and trust them to get the job done.
- Working software is the primary measure of progress and delivery.
- The most efficient and effective method of conveying information to and within a development team is face-to-face conversation with team members and managers.

In Agile development, design and implementation are considered to be the central activities in the software engineering process. Software programming and design and implementation incorporate other activities such as requirements and testing. In an Agile approach, iteration occurs across activities for the requirements and the design developed together, rather than separately. The allocation of requirements and the design planning and development as executed in a series of increments are different from the conventional models. Requirements gathering needs to be completed in order to proceed with the software programming, design, and a development phase as it gives Agile development an extra level of flexibility. An Agile process focuses more on programming code development rather than documentation.

12.2 Agile and Scrum Teams

Software engineers must achieve results from Agile and scrum team meetings. Software engineering leaders actually understand the underlying principles, and they measure things properly to make sure that scrum meetings are held daily to know how everything is going on. If there is a struggle to achieve results from your Agile and scrum teams, the meetings will help you achieve what is needed. Software engineering leaders actually understand the underlying principles of Agile and scrum, and they measure all data and information properly. If there are questions about how to improve specific practices or generally how to improve your Agile journey work with the software engineering Agile leads and you will take many practical ideas to help you take your scrum to the next level. Important Agile adoption for coaching, advice, techniques, and training revolves around the Agile and scrum teams. Software engineering leaders must be the best, and work cases should be performed. It is an important role of software engineering managers and leaders within Agile environments. Examine why those who understand servant leadership know how to effectively support, train, and empower their Agile teams in ways that increase the team performance.

12.3 Leading Agile Software Engineering Teams

Leading an Agile software engineering team requires to provide them coaching, advice, techniques, training, and ways to revolve around the Agile software engineering. Leaders must be the best, and in the worst cases even vilified.

A central and important role of managers and leaders within Agile environments is to explore the patterns of mature Agile performance. Software managers and leaders should know how to effectively support, coach, and empower their Agile teams in ways that increase the team's performance, accountability, and engagement. Exploring training and standards for Agile, situations and guidelines will help trust the software engineering team and when to step in to provide guidance and direction. The software leader's role in Agile at-scale must show good leadership to creating great teams, cultures, and results. In traditional software processes, the role of a software test manager in Agile is to be responsible for all management aspects of their software test team. Agile is self-directed, so software test teams handle all the usual duties as as well as be strategic and look for opportunities.

Following Agile processes will guarantee software engineering teams to be high performers overcoming any challenges to achieve high productivity. Adopting Agile principles, finding typical root causes of those struggles, and acquiring effective behaviors can help drive teams toward greater success.

Scrum isn't the only path to Agile, but it can really help a team to become more in understanding Agile. Agile is all about self-organizing teams by collaboration and what works for software engineering teams with an approach that helps them get started. Agile software development is used to achieve a higher product quality with more predictable on-time delivery to increase customer satisfaction. Self-organizing software engineering teams are a key principle for all Agile methodologies. The truth is that managers and leaders are still very necessary for Agile teams. This chapter guides and provides resources to help software engineering team to embrace an Agile culture, methods, and metrics for transforming into a high-performing powerhouse, and also shows critical ways in which you can provide leadership to your self-directed team.

12.4 Agile Software Engineering Principles

Agile software engineering principles provide a product development framework that maximizes the amount of software engineering planning, programming, and design. A principle will add enough functionality to warrant software engineering releases, and working software is the primary measure of progress.

The Agile software engineering is based on the following principles:

- Customer satisfaction by continuous delivery of software engineering products.
- Deliver working Agile software weekly rather than monthly.
- Software projects that are built by motivated software engineers should be trusted.
- Working software engineering is the primary measure of progress and success.
- Continuous attention to software engineering excellence and programming.
- The best software, requirements, and designs emerge from Agile teams.

A common characteristic of Agile software engineering principles performed is the daily scrum meetings. In scrum meetings, software engineering team members report to each other what they did the previous day and what they intend to do today toward the goals and progress they can see to reach all goals. I always attended the daily scrum meetings conducted for the F-22 Raptor Program to verify the progress.

Agile is one of the popular terms used in software engineering development. Agile development is a different way of managing software engineering development programs and projects. The key Agile principles showing how Agile fundamentally differs from a more traditional Waterfall programming approach in software engineering development are as follows:

- The software engineering teams must be empowered to make decisions.
- Capture requirements at a high level.
- Develop software engineering incremental releases.
- Focus always on frequent delivery of software products.
- Complete each task before moving on to the next task.
- Testing is integrated throughout the software engineering project life cycle.
- A cooperative approach between all software engineering teams is essential.

There are various methodologies and standards that address various aspects of software engineering development principles for the analysis and design of program and project management. Projects and elements of Waterfall development can also be applied in an Agile development approach.

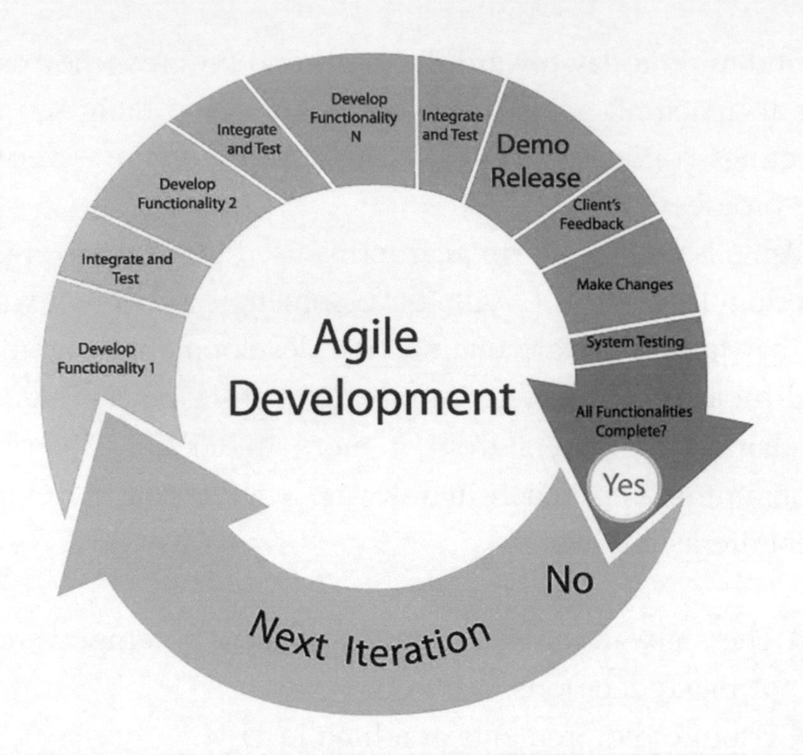

Figure 12.1 Agile development.

Figure 12.1 is a chart showing software engineering principles of Agile development.

Following are the four principles of Agile methodology:

- Agile is all about teamwork and technical excellence.
- Satisfy the program and project customers.
- Welcome change and deliver frequently.
- Work together to build projects.

To create high-performing software engineering teams, Agile methodologies value team members and interactions over processes and tools. All Agile practices seek to increase communication and collaboration through frequent software engineering cycles and consistency of collaborative exchange rooted in a foundation of Agile transparency, trust, respect, and commitment.

Further Reading

Ambler, S., 2002, *Agile Modeling: Effective Practices for EXtreme Programming and the Unified Process*, Hoboken, NJ: John Wiley & Sons. pp. 12, 164, 363. ISBN 978-0-471-20282-0.

Larman, C., 2004, *Agile and Iterative Development: A Manager's Guide*, Boston, MA: Addison-Wesley Professional. p. 253. ISBN 9780131111554. Retrieved 14 October 2013.

Schwaber, K. (2015). Nexus guide. In *2015 7th International Conference on Games and Virtual Worlds for Serious Applications (VS-Games)* (p. 11).

Chapter 13

Software Engineering Testing Results

The software engineering test results are a major factor that demonstrates the quality of products such as software and hardware by running tests during days and nights to ensure that they are working correctly and properly.

13.1 Testing Is Needed

Testing needs to be conducted to assess the quality of the products of software engineering programming and design; hence, testing should be derived from test case methods that can improve the obtained results, thus helping achieve its primary objective.

Developing a software test plan is always critical for software engineering. In some cases, for test plans, the information could be incorporated into a programming and design document. A test plan may be a 200-page document containing test cases and expected results. Test planning begins with the analysis of software engineering problems.

Verification and validation, discussed in Chapter 10, are important requirements and valid operations of software engineering playing quality roles.

Systems engineering defines the role of a software, and analyzes and establishes its function, performance, and validation. The analysis defines the foundation for programming and the design for coding. Testing progresses are important for integration testing.

When programs and project modules are coded, a series of tests are conducted to handle and fix errors in:

- Module interface and connection.
- The algorithm that is implemented by the module.
- Internal data and programming structures.
- Paths through which programs, designs, and projects are working.

Integration testing is a technique for constructing program or project while conducting tests to fix errors during reviews. Once the existing errors are corrected, there could be new ones appear during integration testing. Programs and projects are tested in small segments where errors can be fixed by applying a systematic approach.

13.2 Integration Process

Integration process incorporates all activities directly at each level and structure.

The integration process is performed as following:

- The main control acts as control to testing.
- Tests are conducted at each level that is integrated.
- Completion of each integration process is important.
- Regression testing is conducted to ensure that new errors have been reviewed.

This integration process follows and verifies major control or decision points early in the test process. If major control problems exist, they must be recognized earlier. All input integration processing may be demonstrated before other elements of the projects have been integrated. Having early tasks of functional capability is important for both software engineering activities and customers. The integration tester has the choice of delaying many tests until fixes are replaced with actual modules that perform limited functions to simulate and integrate software engineering from bottom to top. The bottom-up integration strategy is implemented in the following steps:

- Low-level modules are combined to perform software engineering subfunction.
- A control program and project for testing is written for test case input and output.
- Testing of all software engineering integration processes is important.

Selection of integration testing depends on software engineering activities and the time to meet project schedules. Integration follows the pattern illustrated in Figure 13.1 for Teams 1–4 in Periods 1 and 2.

The primary objective of software engineering integration testing is to uncover defects that have been introduced into software programming and design before their implementation. Test cases should be designed to arrest problems that arise in software engineering functions and performance to achieve high quality. The errors identified through software engineering integration process testing must be fixed and corrected, and the software should be made ready for validation. Records of all test results should be maintained as a part of the overall process of software quality assurance for software engineering.

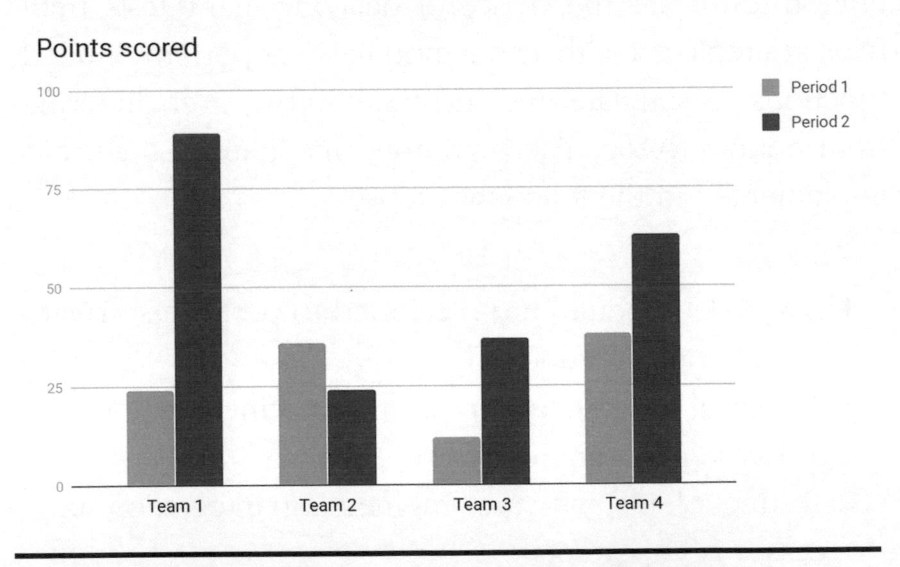

Figure 13.1 Strategy for integration of Periods 1 and 2.

Further Reading

Pressman, R. S., 1988, *Software Engineering: A Beginner's Guide*, New York: McGraw-Hill. ISBN 0-07-050790-2.

Chapter 14

Software Programs and Projects Worked

I have worked in production of software programs and projects for 31 years as a software engineer, software configuration management (SCM), software quality, and technical lead engineer, and I worked with many managers and software engineers to ensure that software programming was performed and brought off according to quality standards. The traditional method of SCM involves identifying software programming/design/development/builds and providing configuration control for software quality engineering support.

Selected software engineering work products and the descriptions to maintain traceability are key points throughout the software engineering life cycle. The Lean concept is the process to compare common information with Agile software development as shown in Figure 14.1.

My software engineering experience was in a common project working toward heavy and important control

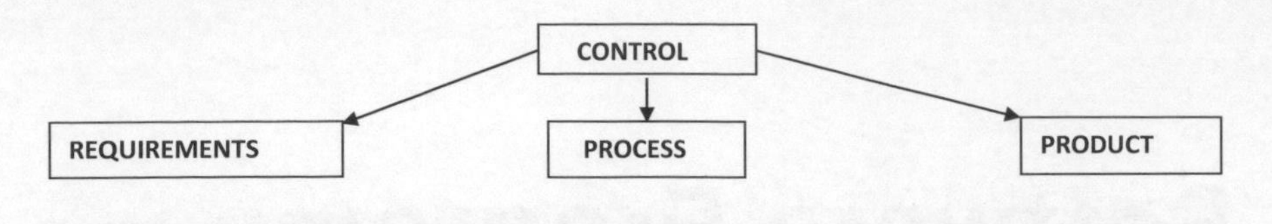

Figure 14.1 Agile management level.

processes. The strategy for software engineering integration provides a road map that describes the steps to be conducted as part of the implementation of software to start integration and delivery activities.

14.1 Military Programs and Projects

I appreciate the 31 years of my experience in working for military programs and projects of the United States of America (USA) and international countries. As mentioned in Chapter 11, I have worked for the B-2 Stealth Bomber and the F-22 Raptor, and I have also given the pictures of the B-2 Stealth Bomber and the F-22 Raptor programs. Other programs I worked are Missile Defense, Airborne Early Warning and Control (AEW&C), Aircrafts for Australia and South Korea, and P-8A Poseidon Navy Aircraft for USA and India. I worked for the P-8A Navy program to support software, quality, and systems engineering. It was great to work for the P-8A Navy program, where I used to think of my father (Louis Summers) every day who was in the Navy during World War II.

Boeing Defense and Space integrated extra crew workstations in order to upgrade Navy P-8A Poseidon maritime surveillance aircraft.

The Navy P-8A has workstations with universal multifunction displays and accommodation for additional workstations and workload sharing.

The P-8A is a militarized version of the Boeing 737 single-aisle jetliner hardened for long-range surveillance, maritime patrol, and anti-submarine warfare missions. The Navy plans to buy 108 P-8A aircrafts from Boeing, which builds the P-8A Poseidon at its factory in Renton, Washington. The Navy also plans to arm the P-8A with the MK 54 torpedo. Working in the P-8A programs was great, and the performance of software and systems engineering was great as well.

I was working in the 1980s for Hercules Missile and Defense as a software engineer supporting software programming and development activities. I worked in many states for Boeing Defense and Space for the military programs and working with companies and the Air Force to ensure software engineering was performed and working for avionics and I worked in Australia and South Korea with the Military Air Force. It was an honor working with the Air Force at many military bases helping them in all software engineering activities. It is important to make the right decisions before delivery of software and hardware end products to the labs and customers. Make sure that software engineering programming, programs, project planning, software requirements, software design, software, and systems integration are successful and quality is maintained in every step of each phase of software engineering design/development life cycle before a delivery of software to customers. It is important to check for the following:

- Software media and data verification and validation are completed.

- Software documentation is released and ready for delivery.
- Necessary first article inspections (FAIs), functional configuration audit (FCAs), and physical configuration audits (PCAs) are conducted, and all action items are closed.

I always worked in FAI, FCA, and PCA with hardware quality engineers to ensure the quality of end products before delivery.

It is crucial to understand and implement the software engineering disciplines during the software engineering programming/design/development life cycle prior to deliveries of software baselines inside software and systems integration environments. Sections of this book define methods for programs, project planning, systems design, software requirements, software design, software implementation, software integration, software and systems integration, subcontractors, delivery, and product evaluations to produce quality end products. Effective methods for software engineering will benefit current and future companies, institutions, military aerospace programs, and/or projects.

It is important in software engineering to conduct system design, software requirements, software design, software test and evaluation, configuration management, quality assurance, and process and product evaluations. Many programs and projects apply processes in Agile, Lean, and Six-Sigma and the American Society for Quality (ASQ).

14.2 Software Engineering Technology

Software engineering technology provides military aerospace and missile defense business solutions through corporate software engineering for supporting automation setup and product and process analysis. Most of software engineering technologies are developed at program and project management, system analysis, application development, data analysis, and engineering. Software engineering and management provide software and product support, technical writing, and effective software development plans.

Software engineering technology services and solutions support the military aerospace and missile defense products in functional testing needs. It helps define and design templates that would best fit organizations for quality and control processes. It was an honor to work in programs for military aerospace and missile defense during my 31 years to ensure all software engineering processes provided the right way for protecting our USA and international countries.

While working for the Boeing Defense and Space programs, I learned so much about the importance of software engineering for aircrafts and missile defense. Many times, I was at Military Air Force bases working software avionics on all the aircrafts ready to fly. It was amazing to be on the AEW&C plane working with pilots to ensure the avionics was working according to the software builds and ready for installation through verification and validation of software quality.

Increasingly technologies for software engineering areimportant for Aircrafts being ready for flight and safety. Most of software quality technology solutions are developed at product and program management, system analysis, application development, data analysis, and software engineering. Software engineering provides system and software support for technical writing and effective software engineering plans. Software engineering technology helps define and design software that would best fit organization processes.

14.3 Software Engineering Communication for Program Plans

Software engineering plans provide communication for software programs and project teams, including the managers, software developers, test analysts, technical writers, functional analysts, and users, to describe the structure of software engineering. Management services are responsible for the process component of software engineering and the evaluation of the product which should be done by joint customer/developer reviews.

The following organizational items should be included or referenced here:

- Software engineering programs and project plans help with communication.
- Program and project team organization charts are used for reviews and approval.

- Software engineering organizational plans and documents include any other teams/organizations participating in software engineering communication processes.

Software engineering communication is an important factor in improvement of companies, institutions, military programs, and successful business processes using plans. Communication using visual goals motivates employees to commit to change by showing expected benefits and successful results.

Many companies, institutions, military programs, and successful businesses are run by management and software engineering technical employees in a technical environment. Often communication for plans for experienced actions that could be comfortable.

Communication is so important in software engineering, and the patterns of communication a software engineer representative uses in programs, projects, companies, institutions, military programs, and successful businesses can determine how to improve their effectiveness.

Communication ultimately determines success and does one of the following:

- Communication and plans solve problems.
- Changes for software engineering teams help better or worse.

Software engineering technology representatives often improve the technologies of software engineering and qualify them for more advanced training of employees. The military aircraft jets built by many companies are a great achievements of software engineering. It was an honor to work in these military programs. My favorite program to work was the F-22 Raptor Program with Lockheed-Martin.

Further Reading

Hockett, M., 2018, *Evaluation Engineering Magazine*, Fort Atkinson, WI. www.linkedin.com/in/mike-hockett-a8332454/.

Kazman, R., Defining the terms architecture, design, and implementation. *Software Engineering Institute*, 2003, Vol 6, No 1, pp. 363–389.

Keller, J., 2019, Military & Aerospace Electronics 2019, http://www.militaryaerospace.com/articles/2013/08/bams-control-complex.html.

Morris, R.A., 2008, *Software Engineering: A Practitioner's Approach*. New York: McGraw-Hill.

Chapter 15

Software Engineering Production

Software engineering contributes to leveraging software production for virtualization and elasticity. Virtualization provides users with access to servers for storage of logical entities that may have direct relation to the underlying physical hardware. For example, a software engineer as a user gets access to carry out several activities in servers, disks, and networking. The benefit of this is that software engineering gets the functionality needed for implementing any changes and maintaining the quality of production.

15.1 Software Engineering Production Elasticity

Software engineering production elasticity refers to the ability to increase virtual resources for production easily and simply. Adding servers involves purchasing, delivering, installing, and configuration, which can happen instantly with virtual servers. Software engineering production elasticity is useful when infrastructure platform services of software engineering have scalability and load issues. For a traditional system and software, the software engineer needs to worry how to increase the user load and other demands on resources that will be handled by software engineering systems. When demands are handled elastically by the platforms, the software engineer has to choose the right platform and has to design within the specifications of that platform.

Software engineers are responsible for good software programming and design, and resource utilization. Software engineering tools can be used for increasing efficiency of software products through test processes. In responsibilities, this is as simple as replacing the software test lab servers and services. The advantages are costs via elasticity and maintenance tools is load leverage elasticity to ramp simulated software demands to world levels. A software engineering test organization could use servers and distribute custom software programming code for its service under test and employ a framework designed for software test products.

The elasticity changes the need for software quality engineers to create product and drive the process to be affected when creating services under tests. The availability of resources and techniques enables an iterative Agile process to be successful. Agile staples, such as continuous integration, are complemented by software engineering technologies such as labs, where a shared infrastructure provides on-demand lab resources throughout the programs and project organizations. Rapid prototyping and flighting are enabled by techniques that are most effective in iterative Agile environments. Software programs, project managers, and software teams will want to consider this natural affinity for Agile when deciding on processes to produce their software services.

The software technologies and applications described illustrate the power services and software tools can bring to the production of software quality engineers. Software engineering developers benefit from the deployment services to access resources for hosting software and data. Software testers should use software tools for applying implementation, loads, and methods for assessing functionality usage. Software teams can employ iterative approaches that better align production of software according to user needs. With such benefits, it appears that elasticity is can be introduced in the software engineering field to produce better software products.

15.2 Software Engineering Production Development

Software engineering production development is based on thinking about the approach that was used in software product development. Many authors describe lean thinking by discussing the practices with underlying principles, and then reapplying those principles to the information in which the authors are working. In this chapter, we will start with the principles that are self-evident in the software engineering world. Always consider the practices that are useful. The foundation of practices is

- Respect software engineers for production development.
- Continuously improve processes and development.
- Respect software engineering knowledge.
- Add value to your customer while retaining ability.

Software engineers need to master multiple skills related to software engineering principles to do production development in general. Developing specific skills helps to be more productive and has an immediate return on investment in the current software engineering role.

An obstacle in day-to-day work of software engineering is lack of technical knowledge, because only technical knowledge can instantly unlock software engineers to make production, gain more in-depth understanding of the technology uses, and update themselves for future development. Working on generic skills set is more easily transferable to other productions and projects. One of the important examples could be learning how to communicate with clients effectively and to be immediately useful to improve valuable paths in software engineering careers. It does matter that projects are built using new technology with software engineers and will still need to communicate with teams and/or customers.

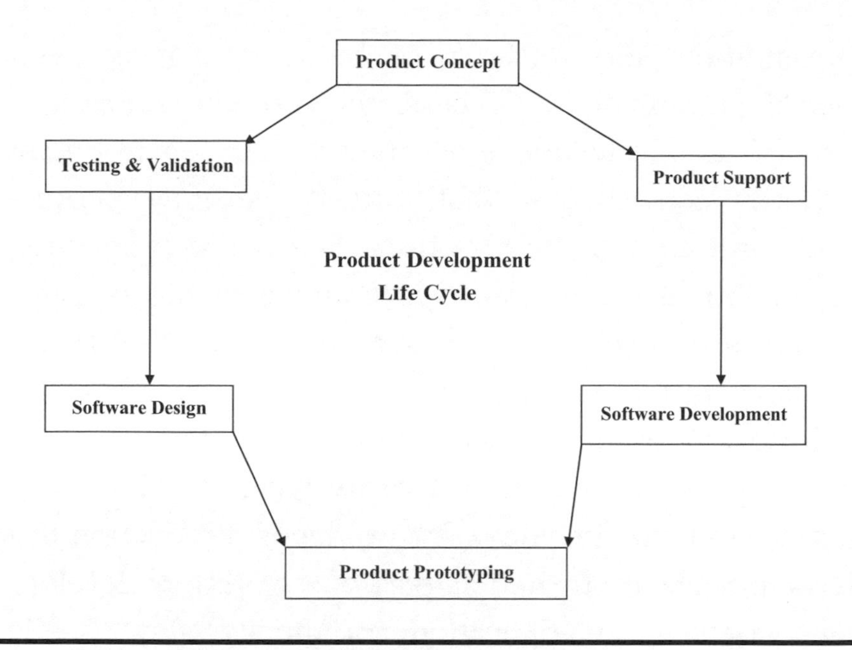

Figure 15.1 Product development life cycle.

I know that production development skills are as generic as software development competencies for all software created to be used, so all software productions need to keep all software engineers in mind. Focusing on all technologies of production development is important to develop software engineering skills for producing more programming and design.

Figure 15.1 shows product development life cycle.

15.3 Software Engineering Production Test Development

A Software Engineering production test development is from the software development environment. Containing special tools and software or being configured with special permissions or access, the test environment is identical to production environment. The environment is closely controlled so that software versions, permissions, and configuration options match the production test environment. Software programmers will use production test development to test changes and enhancements to all custom applications. System administrators will use it to test new versions of software, and users will use it to do unit testing and verify whether all applications meet their specific software needs. Production test developments are treated as part of a workflow developing a website or an application, and the workflow usually includes development, staging, and production. In that case, the workflow might look like this:

- Software developers who work on bugs, features, and updates can be committed directly to the production test development branch.
- Features are implemented and merged into the staging environment for quality assurance and production testing.
- After production testing is complete, they are merged into the development branch.
- The test development branch is merged into production and then deployed to the production environment.

15.4 Software Engineering Delivery

The software engineering delivery approach produces software in short cycles, ensuring that the software can be reliably released at any time manually. It aims at building, testing, and releasing software with a greater speed and frequency. The approach helps reduce the cost, time, and risk of delivering software by allowing for more incremental updates to applications for production. Deployment of a straightforward and repeatable process is important for continuous software engineering delivery.

15.5 Continuous Software Engineering Delivery

Continuous software engineering delivery are similar in their meanings and are often conflated, but they are two different concepts. Development has a broader scope and centers around the cultural change, specifically collaboration of the various teams involved in software engineering delivery from software developers, operations, quality assurance, to management. Continuous software engineering delivery is an approach to automate the delivery process, and it focuses on bringing together different processes and executing them more quickly and more frequently. Thus, development operations can be a product of continuous delivery.

Continuous software engineering delivery is the ability to deliver software that can be deployed at any time through manual releases, in contrast to continuous deployment which uses continuous deployment requirements for delivery.

Software engineering delivery is enabled through the deployment and the purpose of the deployment, which has three components:

- *Visibility* – All aspects of the software engineering delivery system are visible to every member of the team to promote collaboration.
- *Feedback* – Software team members learn about problems as soon as possible when they occur so that they are able to fix them as quickly as possible before delivery.
- *Continually Deploy* – Through a fully automated process, you can deploy and release any version of the software to any environment.

15.6 Software Engineering Productivity Improvement

Software engineering improvement productivity is meant to be a serious job in the grind through your daily software engineering tasks. Enjoying an improvement in productivity gives motivation, provides higher satisfaction, and increases productivity. The happiness reduces stress in software engineering team members so that they can better focus on their work and get more things done, and it also opens up new opportunities for team members to think creatively. The positive encouragement helps increase self-confidence, and publicly applauding the efforts of software engineering team members can be very rewarding.

When starting to work in a company, mostly employees would like to have a friendly, competitive, and encouraging environment and lifestyle. Everyone has to participate for the company to achieve its goals, and there should always be good conversations that help thinking more creatively.

Software engineers fall into this category because they're always looking to see how to improve the software programming and design and how to catch and fix defects. It is important that software engineers understand the power of positive thinking including the benefits in thought process. So software engineers should know how to manage their jobs in their journey to reach the goal. There should be additional activities to maintain positive thinking in all roles of an operation, and they may be perceived in positive actions. Here are some ideas:

- Stay clear on the overall goal of quality that aligns with user needs.
- Software engineers should act as enablers of quality operations.
- Acknowledge if a software engineering work is well done.

- Partake in finding problems being a part of the overall solution.
- Take time to work with the software team members to share their ideas.

Further Reading

Copeland, L. 2012, Better Software, lcopeland@sqe.com, www.TechWell.com BSC&ADC WEST 2012.

Gotimer, O. 1993, Sports Junkie (Go Cuse!). Baseball coach. Doughnut connoisseur. Country music aficionado. Bringing people together.

Chapter 16

CMMI for Software Engineering

The Capability Maturity Model Integration (CMMI) for software engineering collects the best practices to help software team members, companies, and institutions improve process and product development. CMMI always focuses on all software engineering activities for the development of software products and service for meeting the needs of customers and end users. The development of CMMI requires a software engineering team to provide consultation on many project issues and concerns to help improve all software production activities. A CMMI model is shown in Figure 16.1.

16.1 Business and Military Institutions

All companies, institutions, military programs, and successful businesses would always want to provide good support, deliver better services, reduce costs, improve their process, and compliant to standards and specifications.

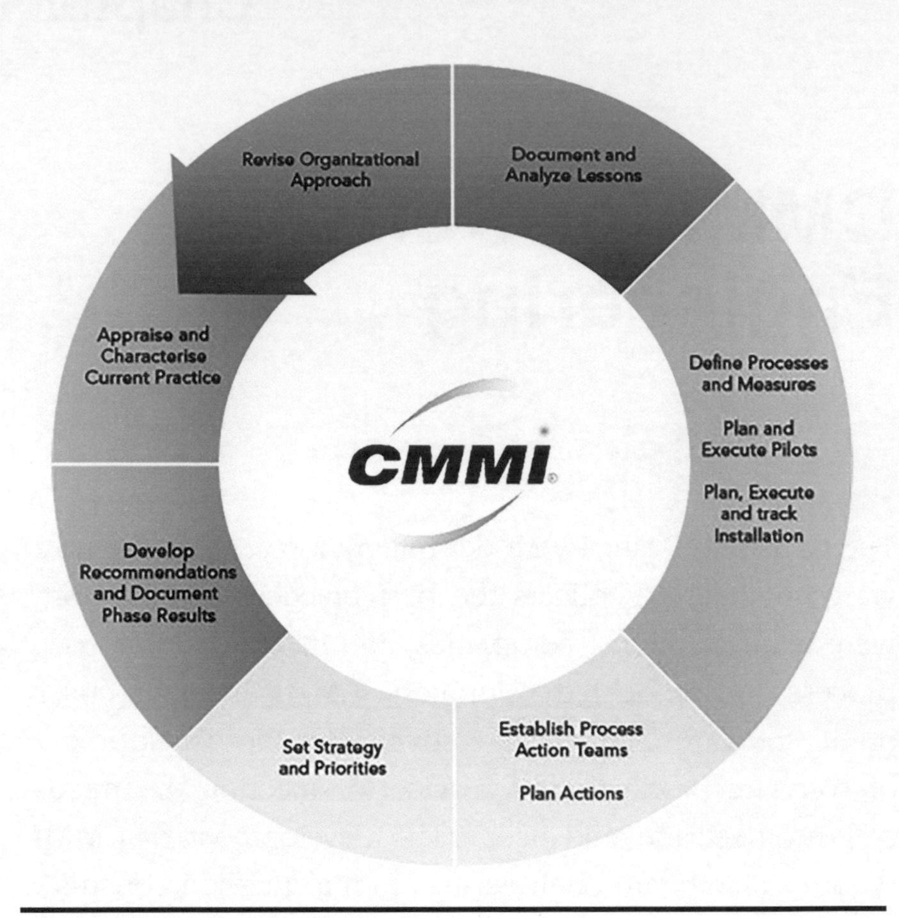

Figure 16.1 A CMMI model.

Management and software team members must be able to manage and control complex maintenance activities. CMMI supports all product services through best practices that address and eliminate numerous issues and problems.

The CMMI model is widely used to implement software engineering processes and quality assurance (QA) in all companies, institutions, military, and business organizations. Process improvement services conducts a research that focuses on all software team members, plans, procedures, and methods to ensure that all tools developed by software engineering are working properly.

The CMMI model for software engineering principles enables continued growth and expansion of software concepts to multiple disciplines and concepts for software engineering to a new level that enables growth and expansion concepts for multiple disciplines. It is designed to help companies, institutions, military, and business organizations improve software engineering products and service development, acquisition, and maintenance processes. CMMI models illustrate continuous representation which focuses on measuring process and development improvement by using capability levels that apply process-improvement achievement to Software Configuration Management (SCM), Software Quality Engineering (SQE) for verification and validation.

16.2 CMMI Framework

CMMI framework has been widely used for many years to embrace software engineering processes such as the development of software and hardware products, service delivery, and purchasing. Depending on the CMMI model framework (CMF) used, constellation of all its process and product areas is shown Table 16.1.

Table 16.1 CMMI Model Framework

Abbreviation	*Name*	*Areas/Places*	*Maturity Level*
REQM	Requirements Management	Software Engineering	2
PMC	Project Monitoring and Control	Project Management	2
PP	Project Planning	Project Management	2
SCM	Software Configuration Management	Support	2
PPQA	Process and Product Quality	Support	2
CPD	Company Process Definition	Process Management	3
MA	Measurement and Analysis	Support	2

16.3 The Purpose of CMMI in Software Engineering

CMMI was developed by the Software Engineering Institute (SEI) at Carnegie Mellon University in 1987 for the purpose of software engineering.

CMMI is used to analyze the approach and techniques followed by any companies or institutions to develop a software engineering product. The guidelines to further enhance the maturity of those software engineering products are based on strong software development practices adopted by the most successful CMMI practices. The CMMI model describes a strategy that should be followed by moving through levels of maturity for process capabilities.

Key process areas (KPAs) define the basic software requirements that should be met by a software process in order to achieve that level of maturity for CMMI. These KPAs form the basis for the management to control software engineering projects; establish a context in which technical methods are applied to produce product models, documents, data, and reports; establish milestones; and ensure better quality.

CMMI focuses on establishing basic management policies for project planning by defining resources required, goals, and constraints of software engineering projects. CMMI presents a detailed plan to be followed systematically for successful completion of a good quality software. Software configuration management focuses on maintaining the performance of the software products, including all its components, throughout their life cycle. Requirements management includes customer reviews and feedback which result in some changes in requirements. Sub-contract management focuses on the effective management of qualified software contractors and the parts of the software that the contractors develop. Software QA guarantees a good quality software product by following certain rules and quality standard guidelines during software development.

16.4 Standard Guidelines for CMMI

The standard guidelines for CMMI take place in well-defined, integrated, project-specific software engineering and management processes. Peer reviews in the CMMI method use a number of methods like walkthroughs, inspections, and software checks. Intergroup coordination is planned interactions between software engineering development teams to ensure efficient and proper fulfillment of customer needs.

Company organization process definition is a key focus on the software development and maintenance of standard software development processes. Organization process focus includes activities and practices that should be followed to improve the process capabilities of an organization. Training programs focus on enhancement of knowledge and skills of team members including the developers and ensuring an increase in work efficiency.

The highest level of software engineering process maturity in CMMI focuses on continuous software process improvement using quantitative feedback. The use of new tools and techniques, and evaluation of software processes prevent the recurrence of software defects. The process change management focuses on the continuous improvement of software processes to improve productivity, quality, and cycle time of software engineering products.

Technology software change management includes use of new technologies to improve product quality and identification of causes of defects to prevent them from recurring in future projects by improving project-defined process.

16.5 Software Engineering with CMMI

Software engineering with CMMI is a set of tools, methods, and practices used to produce software products. The objective of software process management is to produce software products according to plans while simultaneously improving organization's capability to produce better software products with CMMI. In launching a process improvement program, we should first consider the characteristics of a truly effective software engineering process.

The SEI at Carnegie Mellon University incorporates the CMMI to be available for software product and serve as a software engineering process framework that is based on actual practices that reflect the needs of individuals performing software engineering process improvement and software appraisals.

It is important to organize CMMI into the five levels:

- Initial activities for software engineering.
- Managing programs and projects for software engineering.
- Defining capabilities for performance.
- Quantitatively managed concepts.
- Optimizing and prioritizing improvement actions for increasing software engineering process maturity.

Each CMMI process area integration is made up of one or more goals. The goals are targets that are established to ensure control, visibility, and quality. These targets represent

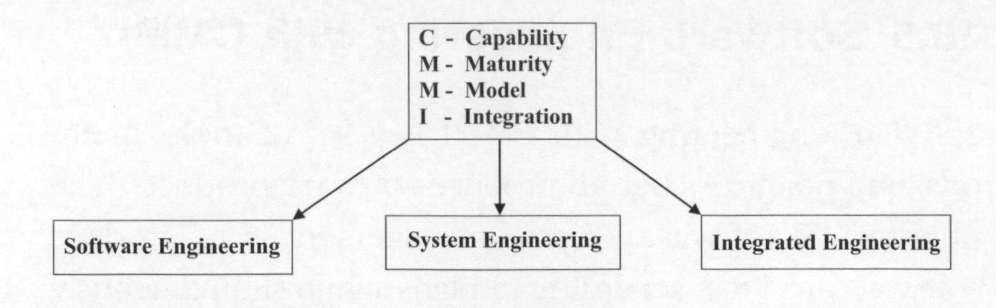

Figure 16.2 Capability Maturity Model Integration.

the real focus of CMMI. The CMMI structure is comprehensive for a collection of process areas with a defined set of goals, and each goal is facilitated by a set of specific practices. It is considered compliant with CMMI if it achieves the goals. To facilitate this effort, CMMI defines a series of practices consistent with each goal of software engineering.

CMMI is shown in Figure 16.2.

Further Reading

Chrissis, M.B., Konrad, M., Shrum, S., 2011. *CMMI for Development: Guidelines for Process Integration and Product Improvement*, Third Edition. Boston, MA: Addison-Wesley Professional. ISBN 978-0-321-71150-2.

Crosby, P.B., 1979, *Quality Is Free the Art of Making Quality Certain.* New York: McGraw-Hill.

Kenett, R.S., Baker, E.R., 2010. *Process Improvement and CMMI for Systems and Software.* CMMI for Development. Boca Raton, FL: Auerbach Publications. www.sei.cmu.edu.

Appendix A: Acronyms and Glossary

Agile: Agile development describes an approach to software development under which requirements and solutions evolve through the collaborative effort of self-organizing and cross-functional teams and their customer/end user.

Audit: An independent examination of a work product for software or a set of work products to assess compliance with specifications, standards, contractual agreements, or other criteria.

Baseline: A specification or product that has been formally reviewed agreed upon and can only be changed through formal control processes.

Build: Operational version of a software product incorporating a specified subset of capabilities that informal and formal work products include in multiple configurations.

Capability Maturity Model Integration: Collection of process models and methods for use in new disciplines to be integrated for organizational structures.

Change Control: The processes by which a change is proposed, evaluated, approved or rejected, scheduled, and tracked.

Change Requests: Requests to software configuration management (SCM) to provide software builds for software systems and use for computer labs and support formal test.

ClearCase: A computer software tool that supports SCM of source code and other software development assets. It also supports design-data management of electronic design artifacts, thus enabling hardware and software co-development.

ClearQuest: Rational ClearQuest is an enterprise-level workflow automation software tool from the Rational Software division. ClearQuest is configured as a bug-tracking system, to track complex processes.

CMMI: Capability Maturity Model Integration collects the best software engineering activities to help all companies, institutions, military programs, and business companies to be successful.

CMMI Framework: Capability Maturity Model Integration Framework guides how CMMI products are developed and integrated, and describes the structure, terminology, and required content of every CMMI model.

Configuration Management: The process of identifying and defining the configuration items in a system, controlling the changes and release of these items throughout the system life cycle, and recording and reporting status of change requests to verify completeness.

C++: A compiled language which is a programming language whose implementations are typically compilers (translators that generate machine mode and software engineering source code and step-by-step executors of source code).

Data: A representation of facts, concepts, or instructions suitable for communication, interpretation, or processing.

Defect: The aspect of software development/coding issues to resolve them on timely basis and drive daily software execution.

Delivery: The point in the product development life cycle at which a product is released to its user for operational use.

Design: The purpose of defining the software architecture, components, modules, interfaces, and data for a software system to satisfy specified requirements.

Development: The process that creates growth, progress, positive change, or the addition of physical, economic, environmental, social, and demographic components.

Engineering Review Board: Established for the software Integrated Product Teams (IPTs) to review and disposition changes that affect controlled software and related documentation.

Effective Process: Improve business to increase expectations, technologies, and competition of effective ways to establish continual and business process improvements.

Elasticity: Changes the need for software quality engineers to create product and drive the process to be affected when creating services under tests.

First Article Inspection: The inspection performed to assure engineering requirements, and processes have been applied to development and release activities.

Hardware: Physical equipment used in data processing, as composed to computer programs, plans, procedures, and associated documentation.

High-Performance Work Team: Teams that drive success though developing and leading high-performance teams to ensure complex tasks are current for the competitive work environment.

Implementation Phase: The period of time in the product life cycle during which work products are created from design documentation.

Inspection: A formal evaluation in which requirements are examined in detail to detect faults, violations of development standards, and other problems.

Integration: Integration in software engineering means combining software parts (so-called subsystems) into one system, and integrated systems demonstrate better performance compared to all independent ones.

Production: The processes and methods used to transform tangible inputs and intangible ideas and information.

Laboratory (Lab): information is a software-based feature that supports software engineering laboratory operations. Key features include data tracking support, flexible architecture, and data exchange interfaces, which fully support software engineering to be ready for delivery.

Life Cycle: The most important parts of software development and requirement analysis are usually carried out by the most skilled and experienced software engineers.

Management: The administration of an organization and business that includes the activities of setting the strategy and coordinating the efforts of employees to management may also refer to those employees that help manage an organization.

Software Metrics: Standard for measurement and evaluation that is used for statistics and to be fixed. A standard of measurement of metrics for software performance, planning work items, and measuring productivity.

Military and Defense: Military and defense is a professional organization formally authorized by sovereign states to use weapons to support the interests of the states and country, and consists of branches such as an Army, Navy, Air Force, Marines, and Coast Guard.

Mission: Driving the growth of management, team members and employees, and all businesses through personal and professional development focused on disciplined execution and quality.

Peer Review: An important part of verification and proven activities for effective defect removal.

Physical Configuration Audit (PCA): Identifies the product baseline for production and acceptance of the work product audited. PCA verifies that the "as-built" configuration correlates with the "as-designed" product configuration and the acceptance test requirements are comprehensive and meet the necessary requirements for acceptance of the production unit.

Policies: A set of policies are principles, rules, and guidelines formulated or adopted by organizations to reach long-term goals and plans.

Procedure: The documented description of a course of action taken to perform activities or resolve problems. Manual steps or processes to be followed.

Process: To perform to defined instructions during the software and product development life cycle.

Prototyping: The activity of creating prototypes of software applications and complete versions of the software program being developed.

Production Test Development: Production Testing Development is not only important but also critical as it allows software testers to detect bugs in real-world scenarios and to ensure if the application works the way is it expected to after the deployment.

Program: A schedule or plan that specifies actions to be taken for companies, institutions, and businesses.

Programming: The process for developing and implementing programming to enable a computer to do certain tasks and computer programming continues to be a necessary process as the internet continues.

Project Plan: A management approach that describes the work to be done, resources required methods to be used, reviews, audits, the configuration management, and quality assurance procedures to be implemented.

Quality: The totality of features and characteristics of a product or service that has the ability to satisfy required needs.

Quality Assurance: A planned and systematic approach to provide adequate confidence that all products conform to established requirements.

Quality Control: The tasks and activities for quality control determine which behaviors are good and add value toward goals knowing that it is important because it helps define risks that organizations can take to help manage organizational support.

Quality Management System: Software industries and software programs who establish, document, implement, and maintain an effective quality management and to continually improve its effectiveness.

Quality Assurance Metrics: Measurement of the degrees to which software possesses given attributes that affect quality and standard for measuring statistics.

Quality Assurance (QA) Plans: QA Plans provide education and learning by ensuring a level of quality in all products and services to be able to build a positive plan for reliability and consistency.

Requirement: A condition or capability needed by a user to solve a problem or achieve an objective. The condition of capability must be met by a system to satisfy a contract, standard, or specification.

Requirement Analysis: The process of studying user needs and arriving at a definition of requirements and verification is also performed.

Requirements Phase: The period of time in the product life cycle during which the requirements of a work product, such as functional and performance capabilities, are defined.

Reviews: Informal or formal review of system requirements, software design, SCM, quality, test, and required data to show compliance to documented plans, processes, and procedures.

Results: To proceed or arise as a consequence, effect, or conclusion.

Risk Management: Process to identify risks and identify an approach to prevent future risks.

Software: Computer programs, procedures, rules, and any documentation pertaining to the operation of data processing systems. It is in contrast to hardware.

Software Configuration Management: Establish and maintain the work product identification process and control changes to identified software work products and their related documentation.

Record and report information needed to manage software work products effectively, including the status of proposed changes and the implementation status of approved changes. Maintain auditable records of all applicable software work products that help verify conformance to specifications, interface control documents, contract requirements, and as-built software configurations.

Software Configuration Management Plan: Configuration Management Plan for controlling and management of software work products during the phase of a software development program.

Software Development Process: The process by which a user needs translated into software requirements, and transformed into design/code being tested, documented, and certified for operational use.

Software Engineering Framework: Frameworks and Methodologies in software engineering can be even more complex under any circumstances because of the issues concerning frameworks, processes, approaches, and methodologies that have been handled in different ways.

Subsystem: It is a part of a computer system used to describe software and usually refers to hardware.

Supplier Data Requirements List: Track Specification Control Documents, Supplier's Design, Approvals, and Acceptance.

Software Engineering: Software Engineering using effective methods is to introduce the notion of software as a product designed and built by software engineers. Software Engineering is important because it is used by a great many people in programs, projects, companies, and institutions. Software Engineers have a moral and ethical responsibility to ensure that the software they design does not cause serious issues and/or problems.

Software Quality: Features and characteristics of a software product that satisfy need and conform to specifications.

Software Quality Assurance: A planned and systematic approach to provide adequate confidence that the product conforms to established requirements.

Software Tools: Computer tools used to develop, test, analyze, and maintain a computer program and its documentation.

Source Code: Computer programs written in a computer language that requires a translation provided by a computer system.

Systems Engineering: Analysis, requirements understanding, and the importance of software design capabilities. Interfaces are defined externally and internally to ensure hardware and software are compatible, supporting team activities.

Subsystem: A group of assemblies or components or both combines to perform a single function.

Suppliers: Supplier requirements establish and maintain documented plans and procedures that define the control of all drawings and documents that relate to the contract or purchase order requirements.

Technology: Applying scientific knowledge to find answers and fix problems used for resources to manufacture software efficiently.

Testing: The process of exercising or evaluating a system by manual or automated means to verify that requirements satisfy expected results.

Test Report: A document describing the conduct and results of testing carried out for a system or system component.

Validation: Validation demonstrates that the product, as provided fulfill its intended use.

Verification: Verification addresses whether the work product properly reflects the specified requirements.

Work Product: A product that consists of requirements, diagrams, documentation, and development folders.

Work Station: A workstation is a special computer designed for technical or scientific applications and is commonly connected to a local area network and run multi-user operating systems.

Index

L

M

O

P